Slow-Wave Substrate-Integrated Waveguide

Slow-Wave Substrate-Integrated Waveguide

Designing of Devices and Antennas

Jing-Ya Deng
School of Physics, Xidian University
China

Jia-Yuan Yin
School of Physics, Xidian University
China

Fengxia Li
School of Physics, Xidian University
China

Yulong Liu
China Research Institute of Radiowave Propagation, Qingdao
China

Li-Xin Guo
School of Physics, Xidian University
China

Xiao-Hua Ma
School of Microelectronics, Xidian University
China

IEEE Press
WILEY

Published by John Wiley & Sons, Inc., Hoboken, New Jersey.
Published simultaneously in Canada.

Library of Congress Cataloging-in-Publication Data Applied for:

Hardback: 9781394314188

Cover Design: Wiley
Cover Image: © oxygen/Getty Images

Set in 9.5/12.5pt STIXTwoText by Straive, Chennai, India

Contents

About the Authors

Jing-Ya Deng (Member, IEEE) received the BS degree in electronical engineering and the PhD degree in electromagnetic field and microwave technology from Xidian University, Xi'an, China, in 2006 and 2011, respectively. From 2011 to 2013, he was a lecturer with the School of Science, Xidian University. From 2013 to 2016, he was an associate professor with the School of Physics and Optoelectronic Engineering, Xidian University, where he has been a full professor since 2016. His current research interests include millimeter-wave antennas, devices, and circuits, multibeam antennas, phased-array antennas, and the high-frequency/very-high-frequency (HF/VHF) antennas design and measurement. Dr. Deng was a recipient of the National Science Foundation for Outstanding Young Scholars of China.

Jia-Yuan Yin received BS degree in electronic science and technology and PhD degree in electromagnetic field and microwave technology from Southeast University, Nanjing, China, in 2013 and 2017, respectively. She was a lecturer before 2022 and is currently an associate professor with the School of Physics, Xidian University. She is mainly engaged in the research of new type of antenna design, metamaterials, including spoof surface plasmon polaritons, metasurface, and the relevant device design.

Fengxia Li received BS in electronic information science and technology from Institutes of Technology of Henan, Henan, China, in 2015, and PhD degrees in microelectronics and solid-state electronics from the University of Electronic Science and Technology of China, Chengdu China, in 2021. She is currently a lecturer with the School of Physics, Xidian University. Her research interests include antenna, artificial electromagnetic material, frequency-selective surfaces, RCS reduction, and wave absorbers.

Yulong Liu received BS in communication engineering from Northwest Polytechnic University, Xi'an, China, in 2003, and the MS in electromagnetic field and microwave technology from China Research Institute of Radiowave Propagation, Qingdao, China, in 2007. Since 2013, he has been a senior engineer with China Research Institute of Radiowave Propagation. His current research interests include the design of HF antennas, phased-array antennas, array signal processing, and directional antenna direction finding.

Li-Xin Guo (Senior Member, IEEE) received the PhD degree in astrometry and celestial mechanics from the Chinese Academy of Sciences, Beijing, China, in 1999. Subsequently, he worked at the Postdoctoral Research Station of "Electromagnetic Field and Microwave Technology" of Xidian University for two years. From April 2001 to May 2002, he was a visiting scholar at the Electromagnetic Wave Laboratory of the College of Electronic Engineering, Kyungpook National University, South Korea. In 1995 and 2007, he visited the University of Rouen in France and the University of Manchester in the United Kingdom for research. He is also an IEEE senior member, a member of the New York Academy of Sciences, a member of the Academic Committee of the Science and Technology Commission of the Ministry of Education, the vice chairman of the Shaanxi Physical Society, a member of the discipline evaluation group of the Shaanxi Provincial Committee on Degrees, a fellow of the Chinese Institute of Electronics, a member of the Electromagnetic Wave Propagation Branch of the Chinese Institute of Electronics, a member of the Electromagnetic Scattering and Inverse Scattering Committee of the Antenna Branch of the Chinese Institute of Electronics, and a member of the Xi'an Municipal Committee of the Chinese People's Political Consultative Conference. He is currently a professor with the school of physics, Xidian University. His current research interests include electromagnetic wave propagation and scattering in complex and random media, computational electromagnetics, inverse scattering, and antenna analysis and design.

Xiao-Hua Ma received the PhD degree in microelectronics and solid-state electronics from Xidian University, Xi'an, China, in 2007. He is currently a professor with the School of Microelectronics, Xidian University, and is also the vice dean of the school. He is the director of the Key Laboratory of Wide Bandgap Semiconductor Materials and Devices of the Ministry of Education. His research interests include microwave devices, enhancement-mode devices, and power electronics devices.

Preface

The performance of multi-beam antennas is crucial for the entire wireless communication system as they can generate multiple high-gain directional beams, each of these beams aimed at different users, thereby improving the communication system capacity. Due to the inclusion of beamforming networks, multi-beam antennas are typically larger in size. With the trend toward miniaturization and integration in wireless communication systems, there is a need for miniaturization and integration design. The substrate-integrated waveguide combines the advantages of microstrip lines and metal waveguides, with low cost, low loss, and easy integration with planar circuits, making it ideal for designing microwave and millimeter-wave multi-beam antennas. At the same time, slow-wave structure is also introduced in the design of antennas and devices, which allows for the additional distributed capacitors and inductors, and can improve the equivalent relative permeability and dielectric constant. This forces the phase velocity of electromagnetic waves on the transmission line to slow down, shortens the waveguide wavelength, and further reduces the size of antennas and microwave devices.

The development of miniaturized and integrated multi-beam antennas using slow-wave substrate-integrated waveguide is vital for enhancing the spectrum efficiency, energy efficiency, and cost efficiency of wireless communication systems. This book provides a comprehensive review and account on the research and analysis on slow-wave substrate-integrated waveguide transmission lines, devices, beamforming networks, and multi-beam antennas. This book will cover topics including fundamentals of a systematic design method and analysis for slow-wave substrate-integrated waveguide, and it is vital to overcome the technical challenges associated with miniaturizing slow-wave microwave devices and antennas. As a rapidly growing frontier field in miniaturized and integrated multi-beam antennas, the topics included in this book will generate significant attention from a variety of research fields and provide guidelines for those who are entering into a related research area or field.

3rd September 2024

Jing-Ya Deng
Xidian University, China

Jia-Yuan Yin
Xidian University, China

Fengxia Li
Xidian University, China

Yulong Liu
China Research Institute of Radiowave Propagation, China

Li-Xin Guo
Xidian University, China

Xiao-Hua Ma
Xidian University, China

Acknowledgments

The authors would like to express their gratitude to the students, reviewers, editors and staff who have made efforts for the publication of this book.

Acronyms

AMC	Artificial magnetic conductor
CPW	Coplanar waveguide
DGS	Defected ground structure
EBG	Electromagnetic bandgap
HMSIW	Half-mode substrate-integrated waveguide
IoE	Internet of Everything
IoP	Internet of People
IoV	Internet of Vehicles
LEO	Low earth orbit
LWA	Leaky-wave antenna
MIMO	Multiple-input multiple-output
MMIC	Monolithic microwave integrated circuit
PEC	Perfect electric conductor
RF	Radio frequency
SIFW	Substrate-integrated folded waveguide
SIW	Substrate-integrated waveguide
SW	Slow wave
SW-ESIW	Slow-wave empty substrate-integrated waveguide
SWF	Slow-wave factor
SWS	Slow-wave structure
SW-SIGGW	Slow-wave substrate-integrated groove gap waveguide
SW-SIW	Slow-wave substrate-integrated waveguide

1 Background

1.1 Research Background

The evolution of wireless communication has experienced significant changes since 1901 [1], particularly with the transition from the 1G–4G Internet of People (IoP) to the current 5G Internet of Everything (IoE) and the anticipated integration of air, earth, and sea communication systems. This transformation has revolutionized communication methods, significantly enhanced communication efficiency, stimulated industrial advancement, and advanced the development of societal information infrastructure. Recent advancements in wireless communication technology have instigated higher demands for channel capacity, integration, and miniaturization of communication systems [2, 3]. Thus, microwave devices are mandated to achieve miniaturization and high integration by reducing circuit size and implementing a planar structure.

Antennas, as devices for receiving and radiating radio signals, play an essential part in the overall effectiveness of wireless communication systems. To improve communication rates, minimize communication delays, and ensure communication quality, it is imperative to enhance the signal-to-noise ratio at user terminals. However, wireless communication is susceptible to obstacles such as building occlusions, multipath fading, and severe weather conditions, leading to a reduction in the signal-to-noise ratio at the user terminal. When increasing the transmitting power is not feasible, high-gain directional antennas can be utilized to enhance the signal-to-noise ratio. Nevertheless, the narrow beamwidth of high-gain antennas restricts their coverage [4]. To address the challenge of reconciling high gain with broad coverage, multi-beam antennas [5–10] can generate multiple high-gain directional beams, each pointing in a different direction, thereby expanding coverage.

According to their technological system, multi-beam antennas may be divided into three types: active digital multi-beam antennas, phased array multi-beam

Slow-Wave Substrate-Integrated Waveguide: Designing of Devices and Antennas, First Edition.
Jing-Ya Deng, Jia-Yuan Yin, Fengxia Li, Yulong Liu, Li-Xin Guo, and Xiao-Hua Ma.

antennas, and passive multi-beam antennas [11]. Phased array multi-beam antennas necessitate numerous adjustable phase shifters, with each antenna unit or subarray requiring connection to a radio frequency (RF) channel. This system, however, is complex and costly [12]. In contrast, active digital multi-beam antennas require intricate beamforming algorithms to achieve phase weighting of signals received or transmitted by each antenna unit in the digital domain, imposing stringent demands on the calculation and processing capabilities of baseband processors [13]. In contrast, passive multi-beam antennas do not necessitate a large number of adjustable phase shifters or complex beamforming algorithms; the beam can be controlled by toggling the RF switch. This approach entails simple processing and low costs. The beamforming network, a critical component of passive multi-beam antennas used to generate multiple high-gain directional beams, includes a typical beamforming network known as the Butler matrix. Only a cascade design of microwave components, such as phase shifters, directional couplers, cross-junctions, and antenna radiation units based on certain topological features, is needed for the Butler matrix. Upon exiting each input port, a high-gain directional beam can be generated correspondingly, offering the benefits of a straightforward structure, easy design, and low loss [14].

The current trend toward miniaturization and integration of modern wireless communication systems necessitates the design of antennas to be in line with the principles of miniaturization and integration. Given that multi-beam antennas require miniaturization due to the larger size of the beamforming network, the Butler matrix—composed of basic microwave components such as couplers, cross-junctions, and phase shifters [15]—can be miniaturized to reduce its physical size, thereby meeting the miniaturization design requirements of the entire multi-beam antenna. In addition, antenna radiation units and beamforming networks might be designed using the substrate-integrated waveguide (SIW), which has been suggested recently, to make it easier to integrate multi-beam antennas with other planar circuits [16]. The SIW combines the benefits of metallic waveguides with microstrip lines, offering low loss, low processing costs, and seamless integration with planar circuits. Furthermore, to further enhance the miniaturization of multi-beam antennas, the introduction of the slow-wave structure (SWS) into antenna and device designs could be beneficial. The SWS shortens the waveguide wavelength, increases the equivalent relative permeability and permittivity, decreases the phase velocity of the electromagnetic wave on the transmission line, and essentially reduces the size of antennas and microwave devices by adding more distributed capacitance and inductance.

Metallic waveguides were first proposed in 1936 [2]; however, their application was limited due to their high machining cost and nonplanar structure, which hindered integration. In 1952, the microstrip transmission line was introduced, enabling the planarization of waveguide structures and improving system integrability. Nevertheless, its semi-open structure resulted in high transmission loss. The gap waveguide structure has become a novel waveguide design in recent decades. It achieves electromagnetic shielding by connecting an artificial magnetic conductor (AMC) and a perfect electric conductor (PEC) in parallel. This novel technology offers low loss, exhibits good electromagnetic compatibility characteristics, and is suitable for the microwave–millimeter-wave field. The gap waveguide exhibits similar electromagnetic characteristics to metallic waveguides and possesses unique advantages compared to traditional metallic waveguides.

Despite its numerous advantages, the gap waveguide is larger than the traditional microstrip circuit, which contradicts the demand for miniaturization in communication systems. While research on SIW, a structure similar to the gap waveguide, has explored various methods for miniaturization, there has been limited research on miniaturizing gap waveguides. Therefore, the realization and miniaturization design of the gap waveguide hold significant research and practical value.

In the context of current mobile communication, achieving high spectral efficiency for 4G and 5G networks is essential. However, various factors, such as noise, multipath fading, and spectrum congestion, significantly affect efficiency improvement. With the increasing demand for miniaturization, single antennas can no longer meet this need, and the available space for antennas has reduced considerably. Consequently, traditional technologies are inadequate. New technologies, such as multiple-input multiple-output (MIMO) technology, have been developed to solve the issue. A key challenge of MIMO technology is enhancing antenna performance within a limited area and mitigating the coupling effect between multiple antenna elements. Therefore, designing MIMO antennas with high isolation in a confined space, by combining antenna miniaturization methods and wideband requirements, is a research problem of significant value that needs to be addressed.

In summary, as a crucial component of wireless communication systems, the performance of microwave–millimeter-wave antennas and devices directly influences the overall system performance. Therefore, researching and designing miniaturized and integrated antennas and devices based on SWS holds great significance in enhancing the spectral efficiency, energy efficiency, and cost efficiency of wireless communication systems.

1.2 Research Status

1.2.1 Slow-Wave Transmission Line

The slow-wave (SW) transmission line is formed by introducing an SWS onto the ordinary transmission line. This introduced SWS enhances the distributed capacitance and inductance effects of the transmission line, effectively increasing the permeability and dielectric constant equivalents of the dielectric substrate. This, in turn, reduces the propagation phase velocity of electromagnetic waves on the transmission line and shortens the guided wavelength, thereby realizing the miniaturization of microwave–millimeter-wave antennas and devices. Hideki Hasegawa of Hokkaido University in Japan was among the first scholars to study the SW propagation mode in microstrip lines in 1971 [17, 18]. Subsequently, extensive research on the SW transmission line was conducted in the academic community. The realization methods of the SW microstrip line include bending line [19] (as illustrated in Figure 1.1a) and loading cycle structure [20, 21] (as illustrated in Figure 1.1b and c). The phase velocity of the electromagnetic wave is decreased by increasing the distributed capacitance and inductance of the transmission line. A group from the South China University of Technology, headed by Chu Qingxin, suggested a mixed-lattice-based SWS in 2012 [22] (see Figure 1.1d). High SW factor (SWF), minimal insertion loss, and linear phase response over a broad frequency range are benefits of the SW microstrip line.

SIW represents an approach to implement microwave and millimeter-wave devices by utilizing metalized via apertures serving as the waveguide's sidewalls and a printed metal layer as the wide wall using printed circuit board (PCB) technology on the upper and lower surfaces of a dielectric substrate. SIW combines the benefits of low losses in metallic waveguides and high integration of planar transmission lines and facilitates the transmission of electromagnetic energy through the regions enclosed by the upper and lower metal surfaces and through holes. Initially proposed by Professor Ke Wu's team at the University of Montreal in 2001 [16], SIW was further examined for its transmission characteristics and design criteria [23]. Subsequently, various microwave devices and antennas based on SIW were investigated [24–29], all of which demonstrated the benefits of strong integration and minimal loss. While SIW dimensions still remained relatively large, efforts have been set forth for further miniaturization. In 2004, the idea of substrate-integrated folded waveguide (SIFW) was first presented by Grigoropoulos et al. from the University of Kent [30]. By folding the common SIW along the centerline of the H-plane, SIFW decreases the transverse width by 50% (Figure 1.2b). This miniaturization technique led to the design of a compact filter based on SIFW [32], followed by a detailed analysis of the design criteria of SIFW by Che et al. from the South China University of Technology [33]. Subsequently, in 2006, Professor HongWei's team from the Southeast University proposed the half-mode SIW (HMSIW) based on the principle of magnetic wall symmetry (Figure 1.2c), which, by reducing the transverse size by 50% and eliminating the need for double-layer plates, contributed to further miniaturization [34]. A comprehensive analysis of HMSIW was conducted by Professor HongWei and his team [35]. While these miniaturization techniques effectively reduced the

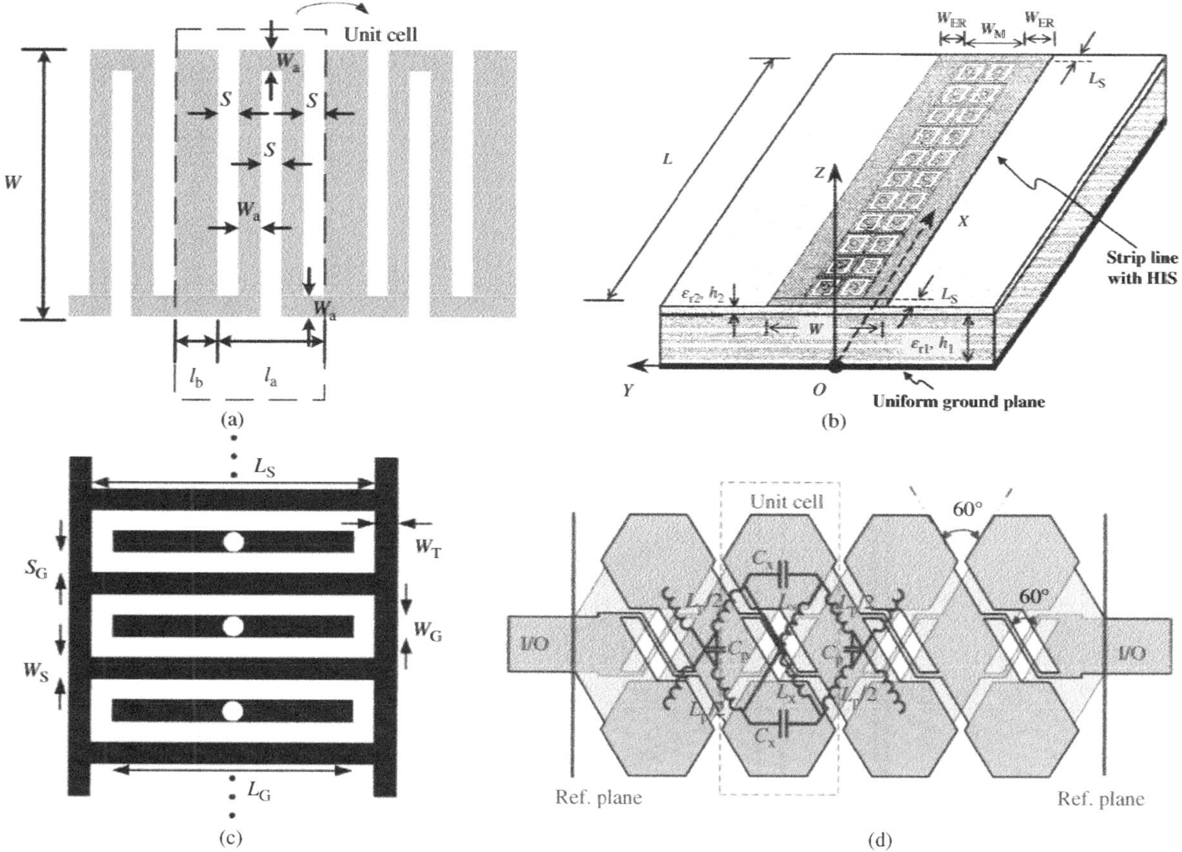

Figure 1.1 Structure diagram of the slow-wave microstrip line. (a) Curved line slow-wave microstrip line. Source: [19]/IEEE. (b) Slow-wave microstrip line loaded with a photonic band gap structure Source: [20]/IEEE. (c) Short-circuited metal-branch-loaded slow-wave microstrip line. Source: [21]/IEEE. (d) Mixed-lattice slow-wave microstrip line. Source: [22]/IEEE.

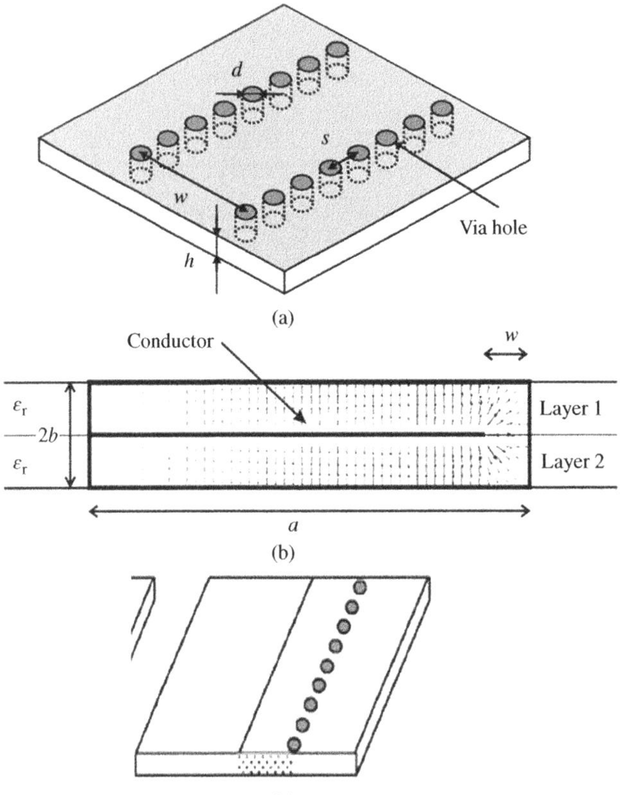

Figure 1.2 Different types of SIW structure diagrams. (a) General SIW structure diagram. Source: [31]/IEEE. (b) Structure of SIFW. (c) HMSIW structure diagram.

transverse dimension of SIW, the reduction of its longitudinal dimension was not addressed.

In 2014, Professor Philippe Ferrari's team at the University of Grenoble introduced SWSs into conventional SIW, resulting in the development of SW substrate-integrated waveguide (SW-SIW) [36]. By loading a metalized blind hole array within the SIW and making electrical contact with the bottom metal floor, the SW-SIW integrated an SWS (Figure 1.3). The concentration of the electric field in the SW-SIW is predominantly between the metalized blind hole and the upper metal floor, which, in comparison to the distance between the upper and lower floors of the SIW, is smaller, consequently increasing the distributed capacitance. Overall, the incorporation of SWSs into SIWs signifies a significant advancement in the miniaturization and performance optimization of SIWs. The equivalent relative dielectric constant increases as a result of the concentration of

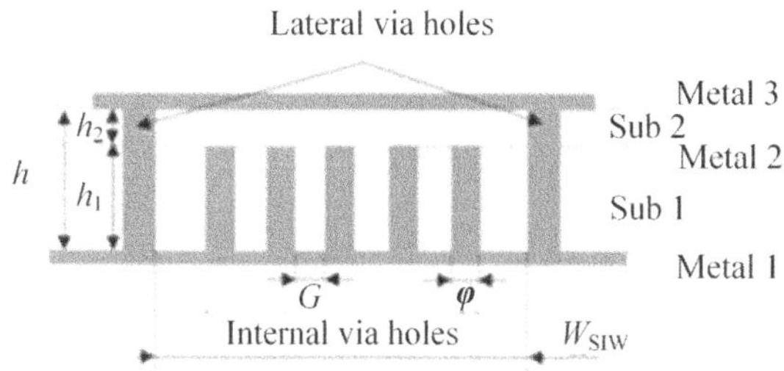

Figure 1.3 Slow-wave SIW loaded by metalized blind holes. Source: [36]/IEEE.

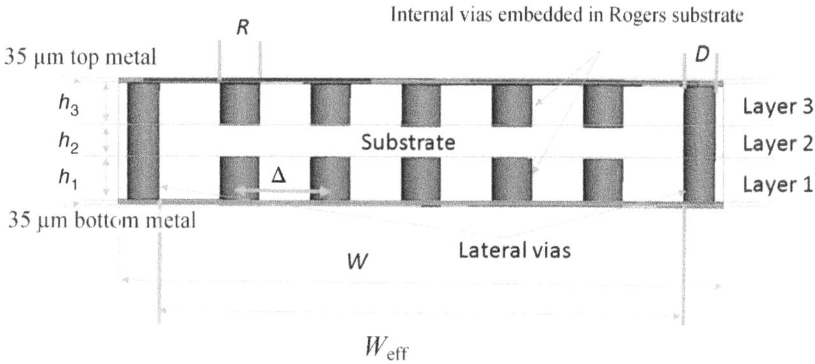

Figure 1.4 Structure diagram of slow-wave SIW loaded with upper and lower symmetric blind holes.

the electric field in the top dielectric plate. As a result, both the cutoff frequency and waveguide wavelength are reduced. When compared with a traditional SIW of the same electrical size, the proposed SW-SIW reduces the transverse and longitudinal dimensions by 40% (see Figure 1.4).

In 2015, Khalil et al. from the Lebanon University proposed an up–down symmetric blind hole SW-SIW based on the aforementioned SW-SIW. This SW-SIW has lower losses and wider bandwidth but also a lower SW coefficient and requires a three-layer dielectric plate. The same year, this research group proposed a half-mode SW-SIW for the first time by loading metal blind holes into an ordinary HMSIW (see Figure 1.5), and the size was reduced by 70% compared with ordinary HMSIW.

In 2016, Professor Wu Ke's team from the University of Montreal proposed an SW-SIW based on multifold lines, which used periodically arranged multifold lines to replace the upper metal floor (as shown in Figure 1.6). The SW-SIW's magnetic field is concentrated below the multifold line, which increases the

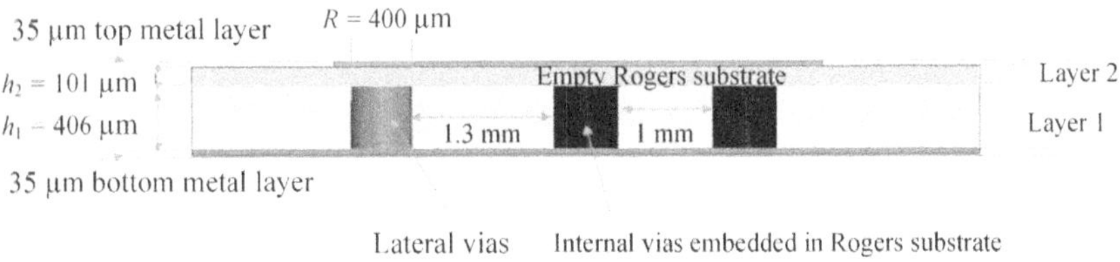

Figure 1.5 Half-mode slow-wave SIW structure. Source: [38]/IEEE.

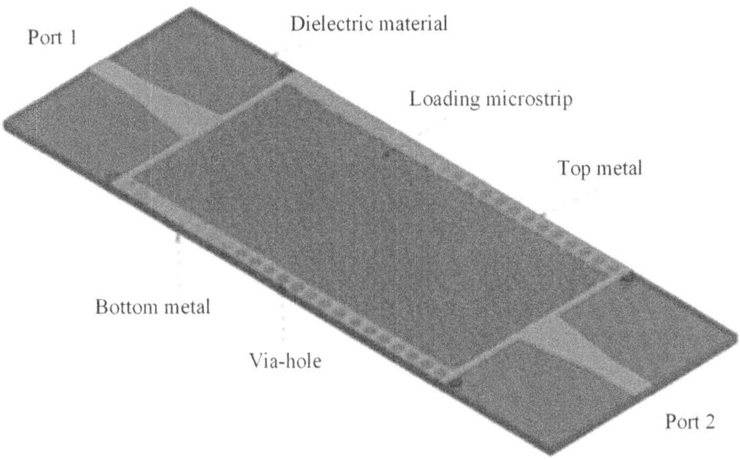

Figure 1.6 Structure diagram of slow-wave SIW loaded with multiple polylines. Source: [39]/IEEE.

distributed inductance and equivalent relative permeability of the SIW. This leads to the downsizing of SIW by lowering the cutoff frequency and waveguide wavelength.

In 2017, Professor Wu Ke's team at the University of Montreal proposed a lumped inductively loaded SW-SIW. By periodically loading lumped inductors on the upper floor (as shown in Figure 1.7), the series inductance and equivalent relative permeability of the SIW are increased, resulting in the size reduction of the ordinary SIW. Additionally, Guan et al. from the University of Electronic Science and Technology of China simulated the corrugation of surface plasmon polariton (SPP) transmission lines by loading metal blind holes inside an ordinary SIW (as shown in Figure 1.8). Due to the low-pass and high-pass features of SPP and SIW, respectively, an SW-SIW based on SPP was proposed, achieving miniaturization and wideband filtering response (see Figure 1.8).

In 2018, Li et al. from the University of California introduced an SW-SIW based on surface plasmon by replacing the upper metal floor with periodically arranged

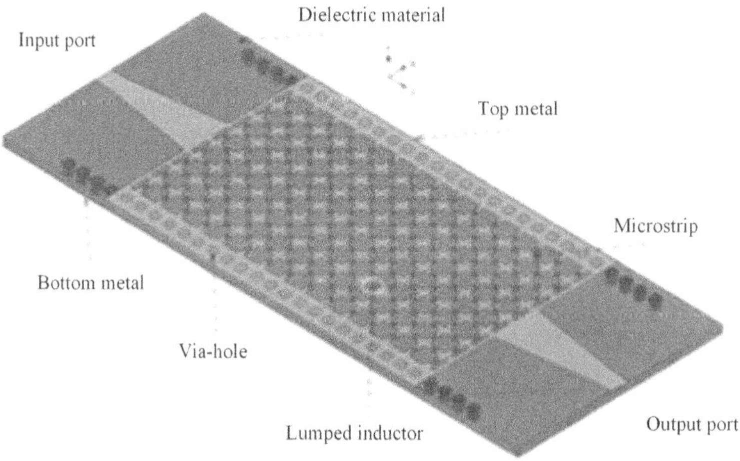

Figure 1.7 Slow-wave SIW structure of lumped inductance loading. Source: [40]/IEEE.

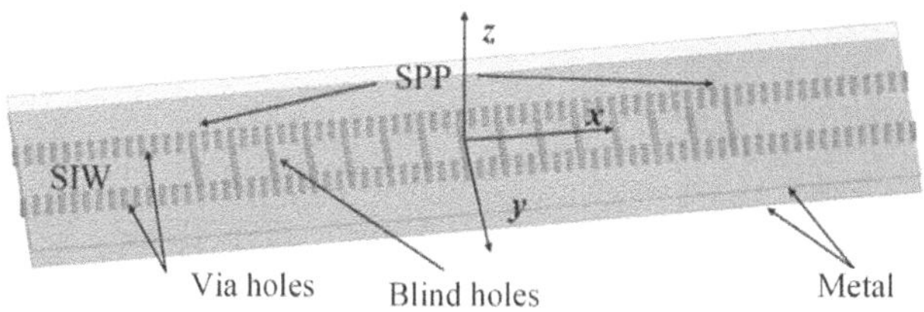

Figure 1.8 Slow-wave SIW based on surface plasmon. Source: [41]/IEEE.

metal patches (see Figure 1.9). The electric field is focused on the top layer of the dielectric plate by connecting the bottom floor and the upper metal patch via a metal hole. It miniaturizes SIW, raises the equivalent relative dielectric constant, and improves the distributed capacitance effect of SIW.

In 2019, Liu et al. from the Nanjing University of Posts and Telecommunications proposed a novel SW-SIW by etching two rows of mirror-symmetric complementary open rings on the upper floor of a common SIW (as shown in Figure 1.10) [43]. This resulted in the concentration of the magnetic field near the complementary open ring, leading to an improvement in the equivalent relative permeability of the SIW and a 20% decrease in the cutoff frequency in comparison to a standard SIW. Based on this SW-SIW, a phase shifter was designed, achieving a 35% reduction in size.

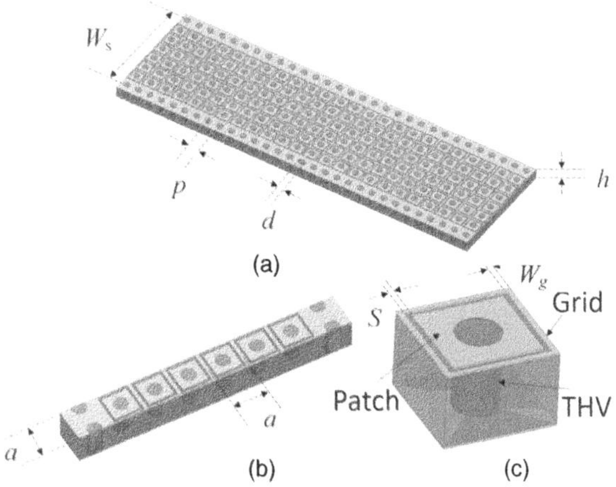

Figure 1.9 Structure diagram of slow-wave SIW based on short-circuit patch loading. (a) Overall structure. (b) A series of Short-circuit patch. (c) Short-circuit patch unit. Source: [42]/IEEE.

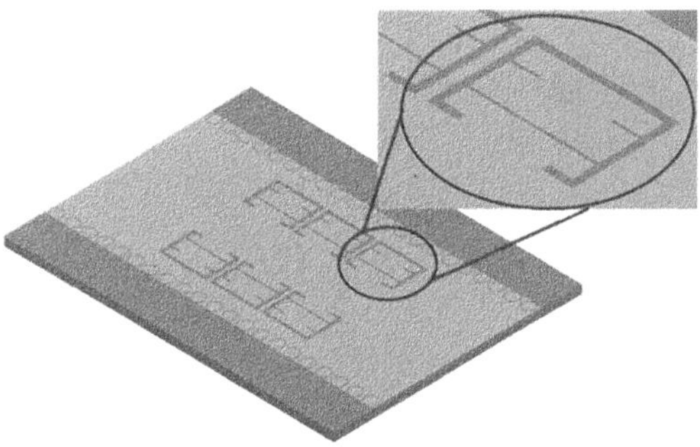

Figure 1.10 Slow-wave SIW loading based on complementary open rings. Source: [43]/IEEE.

In 2020, Zhou et al. from the University of Electronic Science and Technology of China proposed an SW-SIW with asymmetric loading of metalized blind holes, which introduced distributed capacitance in both the upper and lower dielectric substrates (as shown in Figure 1.11) [44]. This modification resulted in an improved equivalent relative dielectric constant and reduced cutoff frequency and

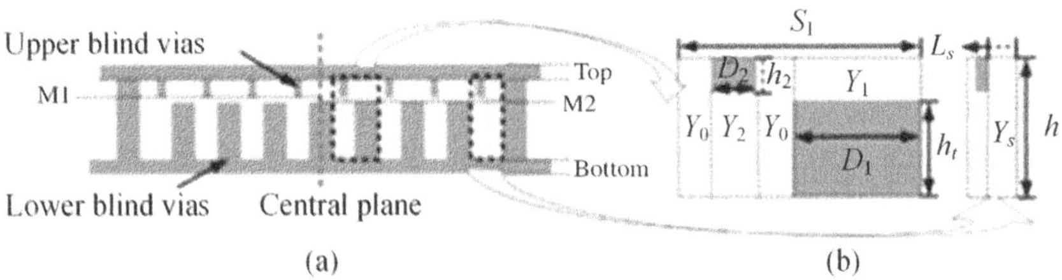

Figure 1.11 (a) Cross section view of slow wave SIW loaded by an asymmetric metal blind hole. (b) Parameters of slow-wave SIW.

phase velocity while simultaneously shortening the transverse and longitudinal dimensions of the SIW.

1.2.2 Slow-Wave Miniaturized Device

Wireless communication systems often utilize various microwave–millimeter-wave devices, such as filters, power splitters, couplers, cross-junctions, and phase shifters. The physical size of these devices is typically related to the operating frequency and waveguide wavelength. SWSs allow these devices to be made smaller by decreasing the phase velocity and shortening the waveguide wavelength without altering the operating frequency.

In 2005, by altering the ground structure on the SIW resonator's surface, Professor Wu Ke was able to decrease the SIW cavity's area (see Figure 1.12). An open-structure eigenvalue issue of SIW cavity structures was thoroughly examined using the finite-difference frequency-domain (FDFD) approach, and the Q factor of such SIW cavities was ascertained. The results showed that high-Q factors are readily attained using the SIW filter. Furthermore, filters made on conventional PCBs have the benefits of compact size, high power capacity, and high-Q factor.

In 2011, there was a proposal for a small SIW filter with a defected ground structure (DGS). Two faulty ground structures were etched into the metal surface of a resonator composed of metalized holes in order to create a third-order filter. One transmission pole was made up of the SIW resonator, while the other two were made up of the two DGS. A coplanar waveguide (CPW) transition structure was used to feed the filter, and the length of the CPW probe cavity was used to modulate the coupling between the source and load. By adjusting the distance between the two defected ground structures, the coupling between the DGS and SIW resonators, was controlled (Figure 1.13).

In 2017, a miniaturized half-mode SW-SIW 3-dB orthogonal coupler based on multifold SW-SIW was designed by Professor Wu Ke from the University of Montreal [47]. The SW coupler had a relative working bandwidth of up to 42%. The amplitude unbalance of the through end and the coupling end in the working band was less than 1 dB, and the phase unbalance was less than 10°. Compared with the common HMSIW 3-dB orthogonal coupler with the same operating frequency, the size was reduced by 40% (Figure 1.14).

In 2018, Bertrand et al. from the Sorbonne University designed a miniaturized fifth-order cavity filter based on an SW-SIW in the form of a metalized blind hole [48] (Figure 1.15). The filter performed similarly to a normal SIW filter operating at the same frequency but with a 68.4% reduction in size. In the same year, Zhou et al. from the University of Electronic Science and Technology of China designed a Wilkinson power divider based on multifold SW-SIW [49]

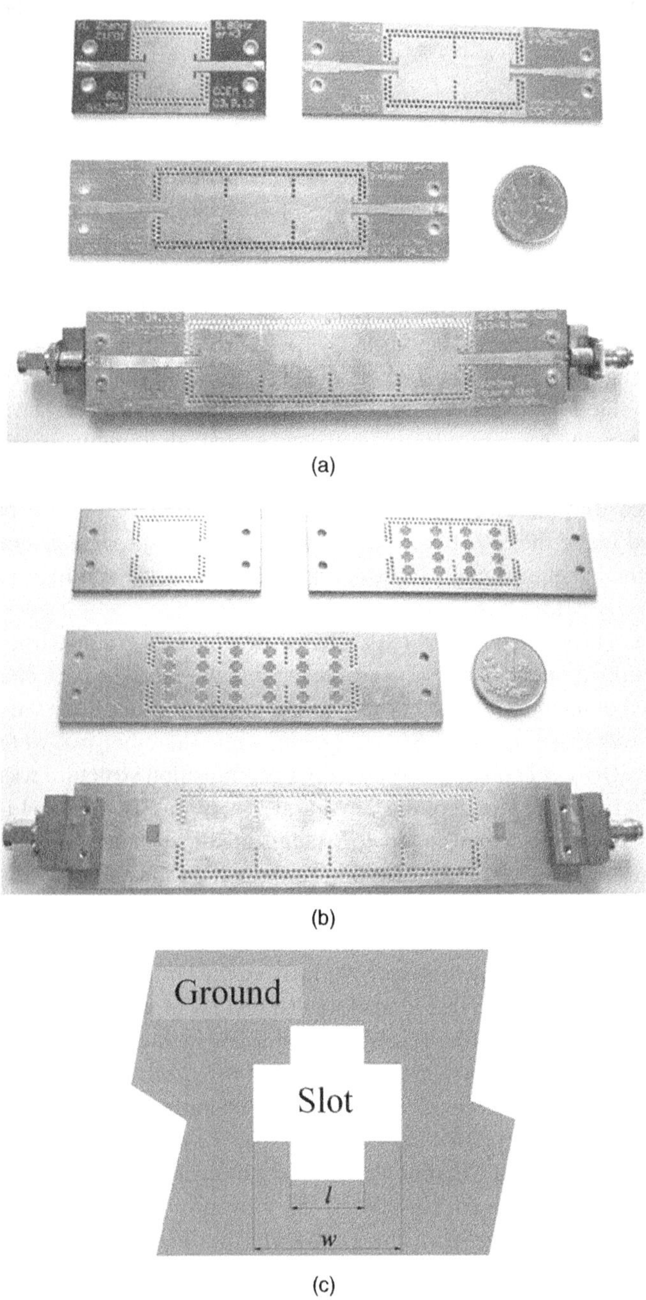

Figure 1.12 Structure of the proposed DGS SIW cavity filters. (a) Top view. (b) Bottom view. (c) Cross-slot. Source: [45]/IEEE.

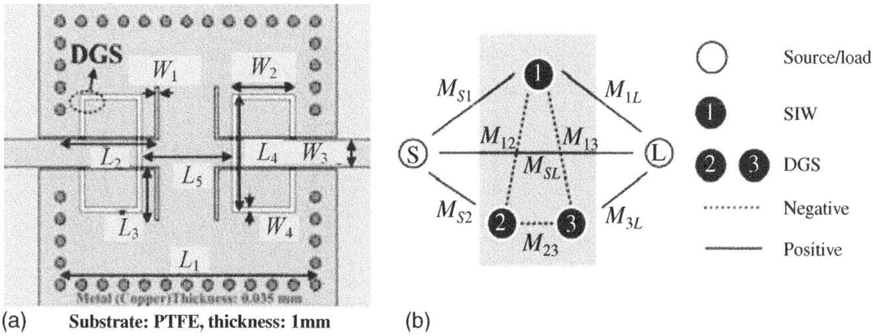

Figure 1.13 (a) Top view and (b) coupling scheme of the proposed filter, respectively. Source: [46]/IEEE.

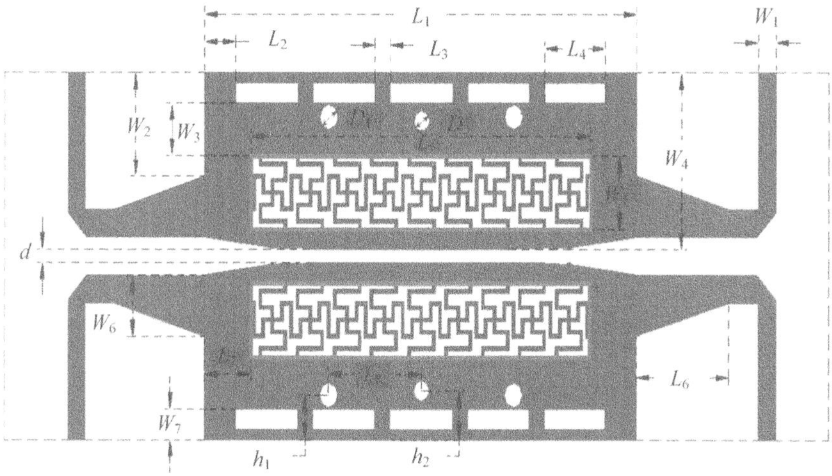

Figure 1.14 Structure diagram of a half-mode slow-wave SIW coupler loaded with multiple lines. Source: [47]/IEEE.

(Figure 1.16). The power divider had a wide relative working bandwidth, and both amplitude and phase unbalances during the working band were less than 1 dB and 8°, respectively, and the output port isolation was better than 10 dB. The size of the power divider under discussion is 40% less than that of a typical SIW power divider running at the same frequency.

In 2019, Bertrand et al. at the Sorbonne University developed a miniaturized 3-dB orthogonal coupler using an SW-SIW loaded with metalized blind holes (refer to Figure 1.17) [50]. By loading metalized blind holes, the coupler's equivalent relative dielectric constant has increased. The SW-SIW coupler has a relative working bandwidth of 18%, a 1-dB amplitude imbalance between the output end

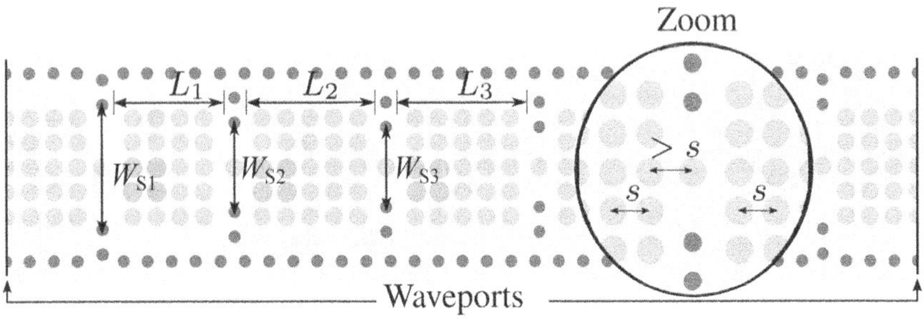

Figure 1.15 Structure diagram of a slow-wave SIW filter loaded with metalized blind holes. Source: [48]/IEEE.

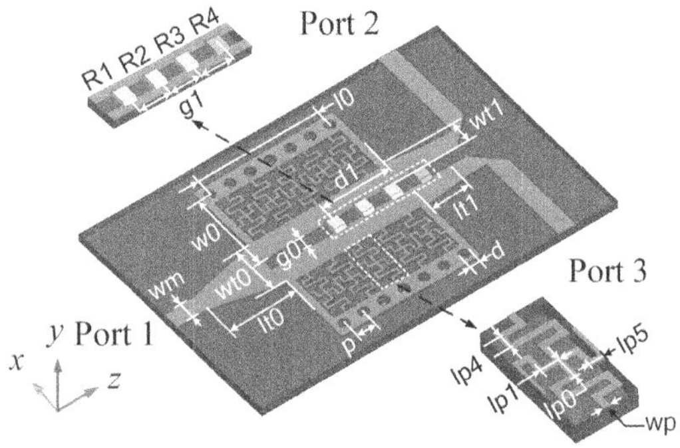

Figure 1.16 Structure diagram of a slow-wave SIW power divider loaded with multiple polylines. Source: [49]/IEEE.

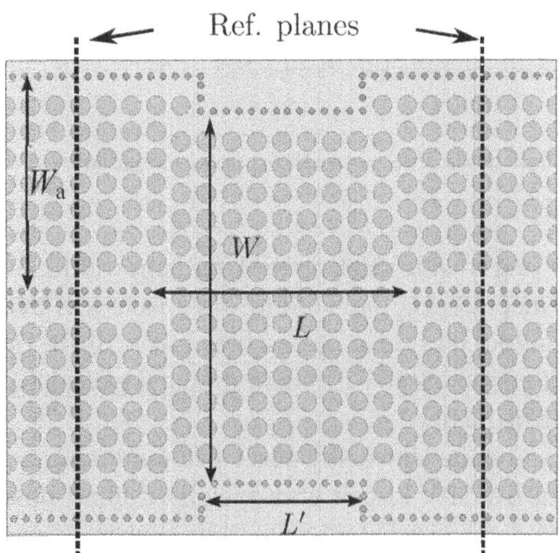

Figure 1.17 Structure diagram of the slow-wave SIW coupler loaded with metalized blind holes. Source: [50]/IEEE.

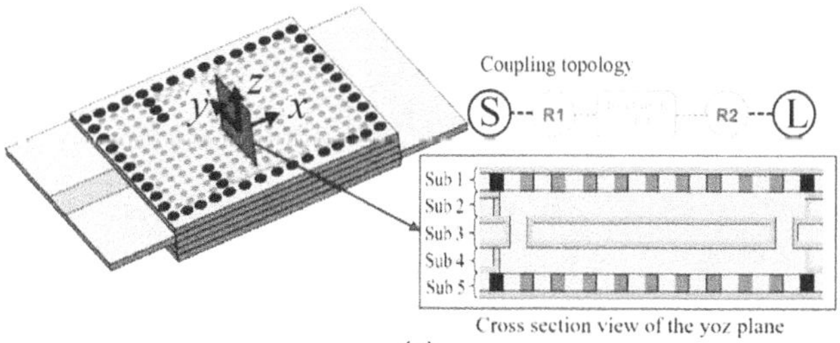

(a)

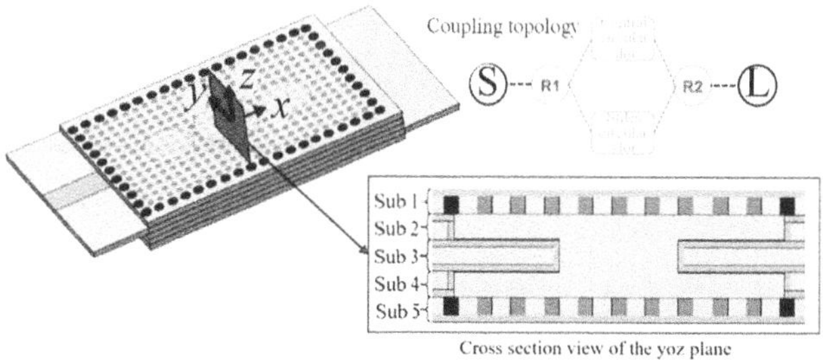

(b)

Figure 1.18 Proposed multilayer SW-ESIW filters with coupling topology. (a) Filter I (magnetic coupling). (b) Filter II (electric coupling). Source: [51]/IEEE.

and the coupling end within the working band, a phase imbalance of only 5°, and an insertion loss of 0.5 dB. This results in a 70% reduction in size compared to a normal SIW coupler at the same operating frequency, with an increase in insertion loss of only 0.4 dB.

In 2021, one electrically linked filter and one magnetically coupled miniaturized filter based on the SW empty SIW (SW-ESIW) were created by Qiang et al. from the Nanjing University of Posts and Telecommunications (see Figure 1.18 [51]). These filters have an insertion loss of <0.8 dB and a relative operational bandwidth of more than 14%. These filters achieve a 35% reduction in size while maintaining filtering performance comparable to ordinary ESIW filters.

1.2.3 Slow-Wave Miniaturized Antenna

In 2011, Chi et al. demonstrated the miniaturization of antennas by periodically loading parallel lumped capacitors [52]. This method efficiently shortens the waveguide wavelength and lowers the phase velocity of electromagnetic waves, representing a concrete implementation of SW miniaturization technology. Utilizing this miniaturization method, the slot-ring antenna (as depicted in Figure 1.19a) and the planar inverted-F antenna (PIFA) (as shown in Figure 1.19b)

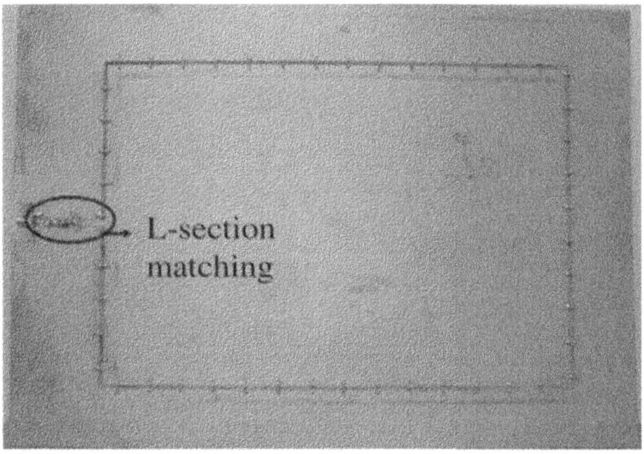

(a)

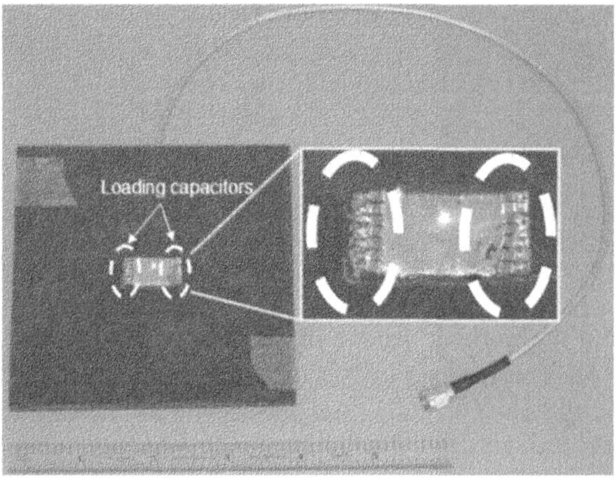

(b)

Figure 1.19 Miniaturized antenna loaded with lumped inductors. Source: [52]/IEEE. (a) Slot-ring antenna [52]. (b) PIFA.

were designed. Both are approximately one-tenth the size of a common slot-ring antenna and PIFA.

In 2014, Elsheakh and other scholars from the Egyptian Institute of Electronics designed an SW miniaturized four-frequency PIFA based on the SW transmission line in the form of periodic bending lines (as shown in Figure 1.20) that have been explored in previous works [53]. The performance of such structures, including the relative working bandwidth and radiation pattern, has been found to be comparable to that of ordinary PIFAs. Despite a reduction of 5–10% in radiation efficiency, the SW PIFA is more than 40% smaller.

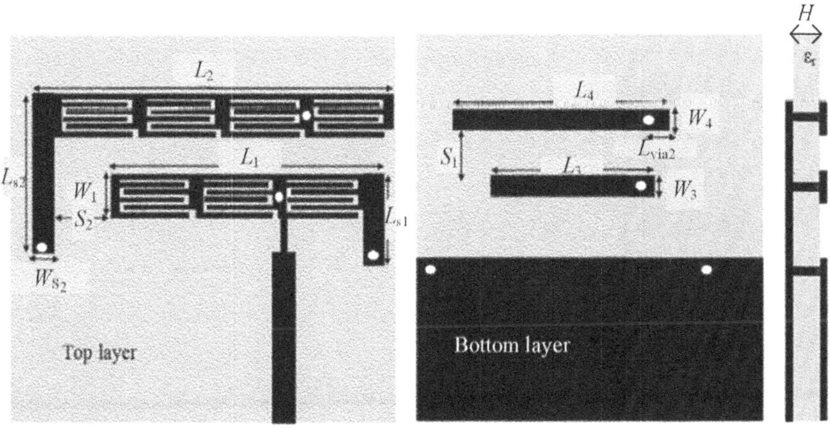

Figure 1.20 Slow-wave miniaturized PIFA loaded with periodic bend lines. Source: [53]/IEEE.

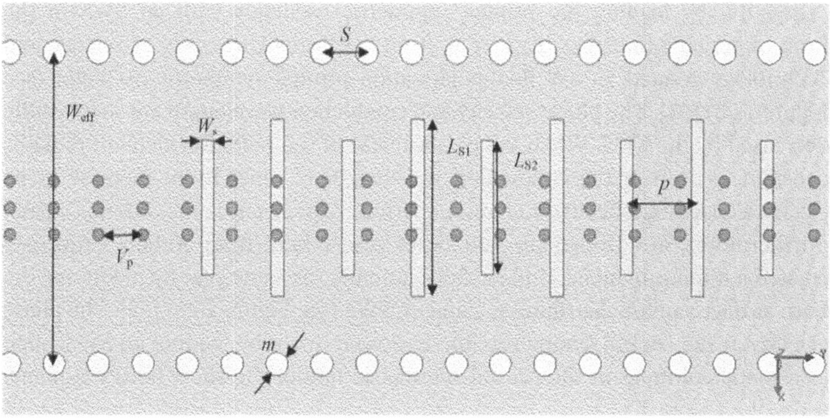

Figure 1.21 Schematic view of the proposed SW-SIW-LWA. Source: Adapted from [54].

Based on the SW-SIW inserted vias [54], Khalim et al. created a small leaky wave antenna (LWA) in 2015 (see Figure 1.21). The SIW-LWA becomes smaller as a consequence of the thin walls created by inserting internal vias into the SIW. This novel design, known as the SW-SIW-LWA, makes use of the vias' SW effect. An adequate gain is maintained despite a 30% reduction in the antenna's transverse plane. Additionally, the SW-SIW-LWA strategy may be a viable option for SIW miniaturization when compared to the miniaturization method using the HMSIW technique presented in the previous papers.

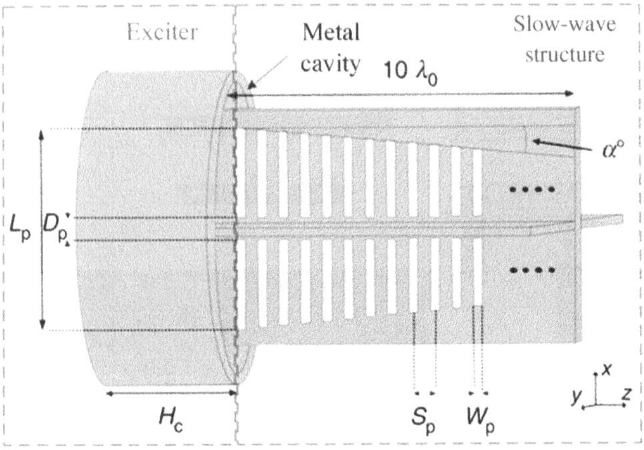

Figure 1.22 Dual-polarization printed oscillator loaded with periodic branches. Source: [55]/IEEE.

In 2017, by loading the printed symmetric oscillator with an SWS in the form of periodic branches, Park et al. from the Korea University of Science and Technology created an SW dual-polarization printed symmetric oscillator (see Figure 1.22) [57]. The phase velocity and conduction wavelength are successfully decreased by the SWS. When the physical size of the radiation element remains constant, a drop in the conduction wavelength is equal to an increase in its electrical size and effective radiation aperture, which raises the antenna's gain. Furthermore, Sun Dongquan and other researchers from Xidian University created a miniature metal H-plane horn antenna the same year by modifying the horn surface's phase distribution using an SWS (see Figure 1.23) [55]. The phase velocity at the central symmetric line is slowed down by loading an SWS, such as a metal column, on the central symmetric line of the short horn's H-plane.

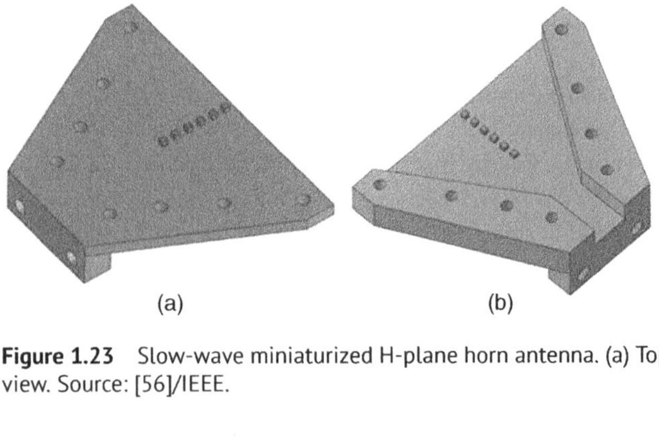

Figure 1.23 Slow-wave miniaturized H-plane horn antenna. (a) Top view. (b) Bottom view. Source: [56]/IEEE.

The horn radiation aperture has a comparatively uniform phase distribution as a consequence of the cylindrical wave within the horn being converted into a plane wave. Lastly, the gain is raised, while the horn's longitudinal length is reduced.

1.2.4 Gap Waveguide

Due to its high electromagnetic and capacitance characteristics and better design convenience, the gap waveguide has been extensively used in millimeter-wave and microwave circuit design. In 2011, Raza et al. utilized a ridge gap waveguide to realize a Ku-band hybrid ring with an operating frequency of 15.3–17.7 GHz (relative bandwidth of 14.5%), with a phase equilibrium of $\pm 5°$ [56]. Additionally, in 2016, Shams proposed an ultra-wideband structure of a ridge gap waveguide power splitter [58]. The reflection coefficient of the power divider during 10.5–20 GHz (95% relative bandwidth) was better than −18 dB.

Microwave–millimeter-wave filters and duplexers with low insertion loss are also important applications for gap waveguides. In 2013, Olmeda et al. demonstrated the use of a vertical polarization ridge gap waveguide, achieving a central frequency of 39.7 GHz for a four-stage diaphragm. The band-pass filter described has a bandwidth of 1 GHz with an insertion loss of 0.42 dB and an in-band echo loss greater than 20 dB [59]. In 2014, Sorkherizi et al. developed a narrow-band X-band band-pass filter using a ridge gap waveguide as an excitation and a groove gap waveguide as the resonant element. The filter has a bandwidth of 72 MHz, a center frequency of 11.6 GHz, an insertion loss of less than 0.6 dB, and a return loss of more than 18 dB [60]. A V-band duplexer based on the groove gap waveguide was created by Rezaee in 2015, with two bands having a frequency separation of 59.5 and 62.5 GHz and a bandwidth of 1 GHz [61].

The gap waveguide is an electromagnetic shield structure based on a wideband noncontact PEC–AMC structure. Building upon this, a noncontact waveguide flange was proposed based on the same noncontact PEC–AMC structure [62, 63] (as shown in Figure 1.24). This noncontact waveguide flange is suitable for millimeter-wave and terahertz rapid measurement systems. The waveguide port to be measured only needs to be aligned with the other half of the noncontact flange and does not require tightening by screws.

The noncontact electromagnetic shield structures used in the gap waveguide also find applications in the encapsulation of millimeter-wave and terahertz circuits. In 2014, Zaman et al. studied the packing properties of the nail-bed structure for the millimeter-wave monolithic microwave integrated circuit (MMIC) amplifier, with experimental results demonstrating effective suppression of cavity resonance within the range of 20–70 dB [64]. Similarly, in 2015, Zhang et al. studied the packaging effect of the nail-bed structure on different microwave–millimeter-wave PCB circuits, with favorable results [65–67].

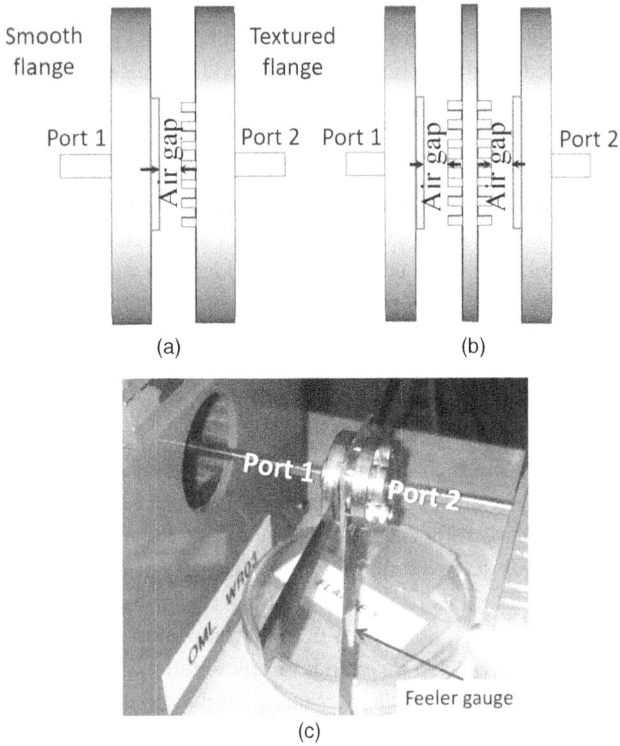

Figure 1.24 (a) Side view of the pin-equipped textured flange attached to a smooth flange. (b) Pin-flange gap adapter with two sides positioned between two flanges. (c) Image of the double-sided pin-flange gap adapter with feeler gauges acting as spacers on both sides, positioned between two flanges. Source: [62]/IEEE.

Additionally, the gap waveguide is especially well suited for terahertz, millimeter-wave, and microwave antenna design due to its low insertion loss. For example, a slot array antenna with a groove gap waveguide as the feed source and a ridge gap waveguide as the input port was suggested by Carrera-Suarez et al. in 2014 [68]. The antenna makes use of waveguides with ridge and groove gaps that are overlaid on the top and bottom layers and connected by a rectangular slot that is inclined. Similarly, an array antenna based on a substrate-integrated ridge gap waveguide and SIW was developed by Raavi et al., and the experimental findings validated the effectiveness of the suggested feeding network [69]. Furthermore, a wideband circularly polarized beam-scanning antenna based on a ridge gap waveguide was shown by Sharkawy et al [70]. It demonstrated a frequency range

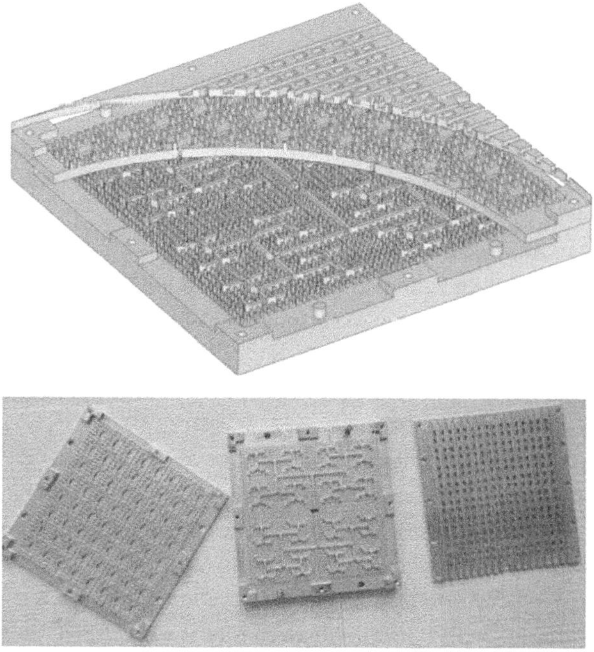

Figure 1.25 Final configuration of the 16 × 16-element array and photograph of the fabricated antenna. The structure has a dimension of 80 × 80 × 12.2 mm. Source: [71]/IEEE.

of 12.5–15.5 GHz (21.4% relative bandwidth), a scanning angle of ±15°, and a gain of 12.5 dBi.

As the PEC plane and perfect magnetic conductor (PMC) plane of the gap waveguide do not need to be in direct physical contact, the feeding network of the array antenna may be merged layer by layer with the antenna's basic construction. The design of array antennas is made much more convenient by this integration. A variety of gap waveguide-based planar high-gain, high-efficiency array antenna types have been described since 2016 [71–75]. As shown in Figure 1.25, one of them, a 16 × 16 planar high-gain array antenna based on a ridge gap waveguide, is presented in Ref. [71]. One layer of radiating components and two layers of gap waveguides make up the antenna. A 16 × 16 rectangular radiation unit makes up the antenna's topmost layer, followed by an 8 × 8 groove gap waveguide cavity resonant layer and an 8 × 8 ridge gap waveguide feeding network layer at the bottom. Operating within the 56–65.5 GHz band, the antenna demonstrates a relative operating bandwidth of 15.6%, a gain exceeding 32.5 dBi, and an efficiency exceeding 70%.

References

1 Balanis, C.A. (2005). *Antenna Theory: Analysis and Design*, 3rd ed., 55–76. New York: Wiley.

2 Pozar, D.M. (2005). *Microwave Engineering*, 4th ed. Wiley.

3 Rao, T.R. (2008). *Introduction to System-on-Package (SOP)*, 1–785. New York: McGraw-Hill Education.

4 Stutzman, W.L. (2024). *Antenna Theory and Design*, 3rd ed. Wiley.

5 Allen, J. (1961). A theoretical limitation on the formation of lossless multiple beams in linear arrays. *IRE Transactions on Antennas and Propagation* 9 (4): 350–352.

6 Stein, S. (1962). On cross coupling in multiple-beam antennas. *IRE Transactions on Antennas and Propagation* 10 (5): 548–557.

7 Dufort, E.C. (1985). Optimum low sidelobe high crossover multiple beam antennas. *IEEE Transactions on Antennas and Propagation* 33 (9): 946–954.

8 Dufort, E.C. (1992). Optimum networks for simultaneous multiple beam antennas. *IEEE Transactions on Antennas and Propagation* 40 (1): 1–7.

9 Lee, W., Kim, J., and Yoon, Y.J. (2011). Compact two-layer rotman lens-fed microstrip antenna array at 24 GHz. *IEEE Transactions on Antennas and Propagation* 59 (2): 460–466.

10 Clark, S., Martin, C., Kolinko, V. et al. (2003). A real-time wide field of view passive millimeter-wave imaging camera//Werner B. *Proceedings of 32nd Applied Imagery Pattern Recognition Workshop*. Washington, DC, USA: IEEE, 250–254.

11 Hong, W., Jiang, Z.H., Yu, C. et al. (2017). Multibeam antenna technologies for 5G wireless communications. *IEEE Transactions on Antennas and Propagation* 65 (12): 6231–6249.

12 Stark, L. (1974). Microwave theory of phased-array antennas-A review. *Proceedings of the IEEE* 62 (12): 1661–1701.

13 Hans, S. (1987). Digital beamforming antennas-An introduction. *Microwave Journal* 30: 107–124.

14 Butler, J. and Lowe, R. (1961). Beamforming matrix simplifies design of electronically scanned antennas. *Electron. Des.* 9: 170–173.

15 Paul, T., Harinath, M., Garg, S.K. et al. (2021). Miniaturized high-power beam steering network using novel nonplanar waveguide butler matrix. *IEEE Microwave and Wireless Components Letters* 31 (6): 678–681.

16 Deslandes, D. and Wu, K. (2001). Integrated microstrip and rectangular waveguide in planar form. *IEEE Microwave and Wireless Components Letters* 11 (2): 68–70.

17 Hasegawa, H., Furukawa, M., and Yanai, H. (1971). Properties of microstrip line on Si-SiO_2 system. *IEEE Transactions on Microwave Theory and Techniques* 19 (11): 869–881.

18 Hasegawa, H., Furukawa, H., and Yanai, H. (1971). Slow wave propagation along a microstrip line on Si-SiO_2 systems. *Proceedings of the IEEE* 59 (2): 297–299.

19 Chang, W.S. and Chang, C.Y. (2012). A high slow-wave factor microstrip structure with simple design formulas and its application to microwave circuit design. *IEEE Transactions on Microwave Theory and Techniques* 60 (11): 3376–3383.

20 Wu, C.K., Wu, H.S., and Tzuang, C.K.C. (2002). Electric-magnetic-electric slow-wave microstrip line and bandpass filter of compressed size. *IEEE Transactions on Microwave Theory and Techniques* 50 (8): 1996–2004.

21 Chang, W.S. and Chang, C.Y. (2011). Novel microstrip periodic structure and its application to microwave filter design. *IEEE Microwave and Wireless Components Letters* 21 (3): 124–126.

22 Huang, J.Q., Chu, Q.X., and Yu, H.Z. (2012). A mixed-lattice slow-wave transmission line. *IEEE Microwave and Wireless Components Letters* 22 (1): 13–15.

23 Bozzi, M., Georgiadis, A., and Wu, K. (2011). Review of substrate-integrated waveguide circuits and antennas. *IET Microwaves, Antennas & Propagation* 5 (8): 909–920.

24 Pasian, M., Silvestri, L., Rave, C. et al. (2017). Substrate-integrated-waveguide E-plane 3-dB power divider/combiner based on resistive layers. *IEEE Transactions on Microwave Theory and Techniques* 65 (5): 1498–1510.

25 Zhou, K., Zhou, C.X., and Wu, W. (2017). Substrate-integrated waveguide dual-mode dual-band bandpass filters with widely controllable bandwidth ratios. *IEEE Transactions on Microwave Theory and Techniques* 65 (10): 3801–3812.

26 Fan, K.K., Hao, Z.C., and Yuan, Q. (2017). A low-profile wideband substrate-integrated waveguide cavity backed E-shaped patch antenna for the Q-LINKPAN applications. *IEEE Transactions on Antennas and Propagation* 65 (11): 5667–5676.

27 Park, W.B., Lee, J.M., Lee, S. et al. (2018). A 18–40 GHz substrate integrated waveguide H-plane horn antenna. *IEEE Transactions on Antennas and Propagation* 66 (11): 6322–6327.

28 Li, H.Y., Xu, J.X., and Zhang, X.Y. (2019). Substrate integrated waveguide filtering rat-race coupler based on orthogonal degenerate modes. *IEEE Transactions on Microwave Theory and Techniques* 67 (1): 140–150.

29 Jamshidi-Zarmehri, H. and Neshati, M.H. (2019). Design and development of high-gain SIW H-plane horn antenna loaded with waveguide, dipole array, and reflector nails using thin substrate. *IEEE Transactions on Antennas and Propagation* 67 (4): 2813–2818.

30 Grigoropoulos, N. and Young, P.R. (2004). Compact folded waveguides//The European Microwave Association. *Proceedings of 34th European Microwave Conference*. Amsterdam, Netherlands: Horizon House Publications Ltd, 973–976.

31 Xu, F. and Wu, K. (2005). Guided-wave and leakage characteristics of substrate integrated waveguide. *IEEE Transactions on Microwave Theory and Techniques* 53 (1): 66–73.

32 Grigoropoulos, N., Sanz-Izquierdo, B., and Young, P.R. (2005). Substrate integrated folded waveguides (SIFW) and filters. *IEEE Microwave and Wireless Components Letters* 15 (12): 829–831.

33 Che, W.Q., Geng, L., Deng, K.A. et al. (2008). Analysis and experiments of compact folded substrate integrated waveguide. *IEEE Transactions on Microwave Theory and Techniques* 56 (1): 88–93.

34 Hong, W., Liu, B., Wang, Y. et al. (2006). Half mode substrate integrated waveguide: A new guided wave structure for microwave and millimeter wave application//IRMMW-THz Committee. *Proceedings of 2006 Joint 31st International Conference on Infrared Millimeter Waves and 14th International Conference on Terahertz Electronics*. Shanghai, China: IEEE, 219–219.

35 Lai, Q.H., Fumeaux, C., Hong, W. et al. (2009). Characterization of the propagation properties of the half mode substrate integrated waveguide. *IEEE Transactions on Microwave Theory and Techniques* 57 (8): 1996–2004.

36 Niembro-Martín, A., Nasserddine, V., Pistono, E. et al. (2014). Slow-wave substrate integrated waveguide. *IEEE Transactions on Microwave Theory and Techniques* 62 (8): 1625–1633.

37 Khalil, M., Kamarei, M., Jomaah, J. et al. (2015). Compact wide-band and low loss substrate integrated waveguide//TAEECE Committee. *Proceedings of 2015 Third International Conference on Technological Advances in Electrical, Electronics and Computer Engineering*. Beirut, Lebanon: IEEE, 130–134.

38 Khalil, M., Kamarei, M., Jomaah, J. et al. (2015). Substrate integrated waveguide miniaturization using slow wave and half mode techniques//Merugu L and Abunjaileh A. *Proceedings of 2015 IEEE MTT-S International Microwave and RF Conference*. Hyderabad, India: IEEE, 173–176.

39 Jin, H.Y., Wang, K.D., Guo, J.P. et al. (2016). Slow-wave effect of substrate integrated waveguide patterned with microstrip polyline. *IEEE Transactions on Microwave Theory and Techniques* 64 (6): 1717–1726.

40 Jin, H.Y., Zhou, Y.L., Huang, Y.M. et al. (2017). Slow-wave propagation properties of substrate integrated waveguide based on anisotropic artificial material. *IEEE Transactions on Antennas and Propagation* 65 (9): 4676–4683.

41 Guan, D.F., You, P., Zhang, Q.F. et al. (2017). Hybrid spoof surface plasmon polariton and substrate integrated waveguide transmission line and its application in filter. *IEEE Transactions on Microwave Theory and Techniques* 65 (12): 4925–4932.

42 Li, X., Dhwaj, K., Itoh, T. (2018). Single-layer slow-wave substrate integrated waveguide with enhanced capacitance//Kanamaluru S. *Proceedings of 2018 IEEE/MTT-S International Microwave Symposium*. Philadelphia, PA, USA: IEEE, 316–318.

43 Liu, S. and Xu, F. (2019). Novel substrate-integrated waveguide phase shifter and its application to six-port junction. *IEEE Transactions on Microwave Theory and Techniques* 1–8.

44 Zhou, Y.L., Jin, H.Y., Huang, Y.M. et al. (2020). Slow-wave effect enhanced substrate integrated waveguide with multi-antipodal blind via-holes and distributed metal strips. *IEEE Microwave and Wireless Components Letters* 30 (8): 753–756.

45 Zhang, Y.L., Hong, W., Wu, K. et al. (2005). Novel substrate integrate waveguide cavity filter with defected ground structure. *IEEE Transactions on Microwave Theory and Techniques* 53 (4): 1280–1287.

46 Shen, W., Yin, W., and Sun, X. (2011). Compact substrate integrated waveguide (SIW) filter with defected ground structure. *IEEE Microwave and Wireless Components Letters* 21 (2): 83–85.

47 Jin, H.Y., Zhou, Y.L., Huang, Y.M. et al. (2017). Miniaturized broadband coupler made of slow-wave half mode substrate integrated waveguide. *IEEE Microwave and Wireless Components Letters* 27 (2): 132–134.

48 Bertrand, M., Pistono, E., Kaddour, D. et al. (2018). A filter synthesis procedure for slow wave substrate integrated waveguide based on a distribution of blind via holes. *IEEE Transactions on Microwave Theory and Techniques* 66 (6): 3019–3027.

49 Zhou, Y.L., Huang, Y.M., Jin, H.Y. et al. (2018). Slow-wave half-mode substrate integrated waveguide 3-dB wilkinson power divider/combiner incorporating nonperiodic patterning. *IEEE Microwave and Wireless Components Letters* 28 (9): 765–767.

50 Bertrand, M., Dirani, H.E., Pistono, E. et al. (2019). A 3-dB coupler in slow wave substrate integrated waveguide technology. *IEEE Microwave and Wireless Components Letters* 29 (4): 270–272.

51 Qiang, J.X., Xu, F., Yang, L. et al. (2021). Enhanced performance of multilayer bandpass filter using slow wave empty substrate integrated waveguide (SW-ESIW). *IEEE Microwave and Wireless Components Letters* 31 (12): 1279–1282.

52 Chi, P.L., Waterhouse, R., and Itoh, T. (2011). Antenna miniaturization using slow wave enhancement factor from loaded transmission line models. *IEEE Transactions on Antennas and Propagation* 59 (1): 48–57.

53 Elsheakh, D.M. and Safwat, A.M.E. (2014). Slow-wave quad-band printed inverted-F antenna (IFA). *IEEE Transactions on Antennas and Propagation* 62 (8): 4396–4401.

54 Khalil, M., Kamarei, M., Jomaah, J., and Ayad, H. (2015). Compact SIW leaky wave antenna. *2015 Third International Conference on Technological Advances in Electrical, Electronics and Computer Engineering (TAEECE)*, Beirut, Lebanon, 124–129.

55 Sun, D.Q. and Xu, J.P. (2017). Compact phase corrected H-plane horn antenna using slow-wave structures. *IEEE Antennas and Wireless Propagation Letters.* 16: 1032–1035.

56 Raza, H. and Yang, J. (2011). A low loss rat race balun in gap waveguide technology. *Proceedings of the 5th European Conference on Antennas and Propagation (EUCAP)*, 1230–1232.

57 Park, S.B. and Eom, S.Y. (2017). Printed cross-dipole antenna with slowwave structure for highly efficient and directive WLAN applications [J]. *Electronics Letters.* 53 (13): 826–828.

58 Shams, S.I. and Kishk, A.A. (2016). Wide band power divider based on ridge gap waveguide. *2016 17th International Symposium on Antenna Technology and Applied Electromagnetics (ANTEM)*, 1–2.

59 del Olmo-Olmeda, A., Baquero-Escudero, M., and Boria-Esbert, V.E. et al. (2013). A novel band-pass filter topology for millimeter-wave applications based on groove gap waveguide. *2013 IEEE MTT-S International Microwave Symposium Digest (MTT)*, 1–4.

60 Sorkherizi, M.S., Khaleghi, A., and Kildal, P.-S. (2014). Direct-coupled cavity filter in ridge gap waveguide. *IEEE Transactions on Components, Packaging and Manufacturing Technology* 4 (3): 490–495.

61 Rezaee, M., Zaman, A.U., and Kildal, P.S. (2015). V-band groove gap waveguide diplexer. *2015 9th European Conference on Antennas and Propagation (EUCAP)*, 1–4.

62 Rahiminejad, S., Pucci, E., Vassilev, V. et al. (2016). Polymer gap adapter for contactless, robust, and fast measurements at 220–325 GHz. *Journal of Microelectromechanical Systems* 25 (1): 160–169.

63 Rahiminejad, S., Pucci, E., and Haasl, S. (2014). Micromachined contactless pin-flange adapter for robust high-frequency measurements[I]. *Journal of Micromechanics and Microengineering* 24 (8): 1–12.

64 Zaman, A.U., Alexanderson, M., and Kildal, P.-S. (2014). Gap waveguide PMC packaging for improved isolation of circuit components in high-frequency microwave modules. *IEEE Transactions on Components, Packaging and Manufacturing Technology* 4 (1): 1075–1084.

65 Zhang, J., Zhang, X., Shen, D. et al. (2015). Gap waveguide-based PMC packaging for via holes-caused nonsmooth PEC surface. *IEEE Transactions on Components, Packaging and Manufacturing Technology* 5 (12): 1828–1838.

66 Zhang, J., Zhang, X., Shen, D. et al. (2016). Gap waveguide PMC packaging for a SIW-GCPW-based filter[J]. *IEEE Microwave and Wireless Components Letters* 26 (3): 159–161.

67 Zhang, J., Zhang, X., and Shen, D. (2016). Gap waveguide PMC packaging for two-layer PEC surfaces. *IEEE, Transactions on Components, Packaging and Manufacturing Technology* 6 (6): 906–916.

68 Carrera-Suárez, L.F., Navarro-Méndez, D.V., and Baquero-Escudero, M. et al. (2014). A novel twist between Gap Waveguides for compact slot-array antennas. *2014 IEEE Antennas and Propagation Society International Symposium (APSURS)*, 456–457.

69 Razavi, S.A., Kildal, P.S., and Xiang, L. et al. (2014). Design of 60 GHz planar array antennas using PCB based microstrip-ridge gap waveguide and SIW. *The 8th European Conference on Antennas and Propagation (EuCAP)*, 1825–1828.

70 Sharkawy, M.A. and Kishk, A.A. (2014). Wideband beam-scanning circularly polarized inclined slots using ridge gap waveguide. *IEEE Antennas and Wireless Propagation Letters* 13: 1187–1190.

71 Zarifi, D., Farahbakhsh, A., and Zaman, A.U. (2016). Design and fabrication of a high-gain60-GHz corrugated slot antenna array with ridge gap waveguide distribution layer. *IEEE Transactions on Antennas and Propagation* 64 (7): 2905–2913.

72 Vosoogh, A. and Kildal, P.-S. (2016). Corporate-fed planar 60-GHz slot array made of three unconnected metal layers using AMC pin surface for the gap waveguide. *IEEE Antennas and Wireless Propagation Letters* 15: 1935–1938.

73 Vosoogh, A., Kildal, P.S., and Vassilev, V. (2016). A multi-layer gap waveguide array antenna suitable for manufactured by die-sink EDM. *2016 10th European Conference on Antennas and Propagation (EuCAP)*, 1–4.

74 Farahbakhsh, A., Zarif, D., and Zaman, A.U. et al. (2016). Corporate distribution networks for slot array antenna based on groove gap waveguide technology. *2016 10th European Conference on Antennas and Propagation (EuCAP)*, 1–3.

75 Cao, B., Wang, H., and Huang, Y. (2016). W-band high-gain TE_{220}-mode slot antenna array with gap waveguide feeding network[J]. *IEEE Antennas and Wireless Propagation Letters* 15: 988–991.

2

Design and Operating Principle of SW-SIW

2.1 Fundamentals of Slow-Wave Transmission Lines

Slow wave (SW) refers to an electromagnetic (EM) wave with a phase velocity that is less rapid than light. SW transmission lines are formed by incorporating SW structures (SWSs) into conventional transmission lines. The phase velocity of an EM wave can be expressed as:

$$v_p = \lambda f \tag{2.1}$$

where λ and f denote the wavelength and frequency of the EM wave, respectively. By integrating the SWSs, the phase velocity v_p and wavelength λ are reduced at a fixed frequency f. This reduction enables the miniaturization of microwave components and antennas, given that their physical dimensions are related to the wavelength.

A generalized transmission line model (illustrated in Figure 2.1) can be utilized to analyze SW transmission lines. According to Kirchhoff's law, the voltage $V(z)$ and the current $I(z)$ at any point on the transmission line can be determined as:

$$\frac{dV(z)}{dz} = -(R + j\omega L)I(z) \tag{2.2}$$

$$\frac{dI(z)}{dz} = -(G + j\omega C)V(z) \tag{2.3}$$

The voltage $V(z)$ and the current $I(z)$ can then be derived by separating variables:

$$\frac{d^2V(z)}{dz^2} = \gamma^2 V(z) \tag{2.4}$$

$$\frac{d^2I(z)}{dz^2} = \gamma^2 I(z) \tag{2.5}$$

$$\gamma = \alpha + j\beta = \sqrt{(R + j\omega L)(G + j\omega C)} \tag{2.6}$$

Slow-Wave Substrate-Integrated Waveguide: Designing of Devices and Antennas, First Edition.
Jing-Ya Deng, Jia-Yuan Yin, Fengxia Li, Yulong Liu, Li-Xin Guo, and Xiao-Hua Ma.

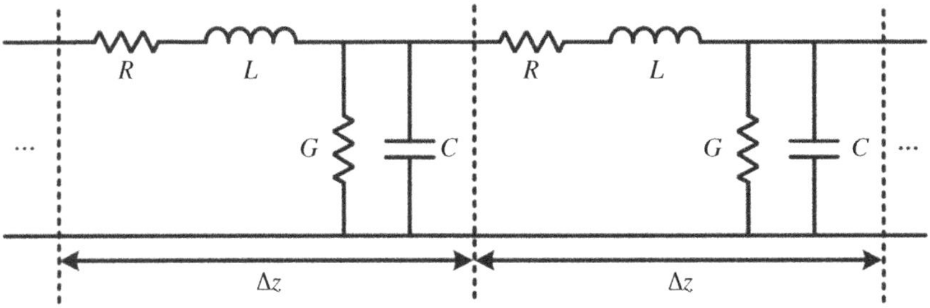

Figure 2.1 Generalized transmission line model.

where γ represents the complex propagation constant, the real part α denotes the transmission line loss, and the imaginary part β defines the phase shift per unit length. For low-loss transmission lines, the conditions $R \ll L$ and $G \ll C$ hold and can be ignored. Thus, the phase velocity v_p and phase shift constant can be determined as:

$$\beta = \omega\sqrt{LC} \tag{2.7}$$

$$v_p = \frac{\omega}{\beta} = \frac{1}{\sqrt{LC}} = \lambda f \tag{2.8}$$

The phase velocity v_p is inversely proportional to the distributed capacitance C and inductance L, as shown in Eq. (2.8). By loading the SWSs, SW technology aims to decrease the EM wave's phase velocity by increasing the transmission line's distributed capacitance and inductance. As a result, microwave components and antennas may have smaller physical sizes at a particular working frequency. However, because of the transmission line's characteristic impedance, raising the distributed capacitance or inductance alone would cause an impedance mismatch between the SW-loaded and non-SW-loaded units, which would lower the transmission line's operational bandwidth.

The conventional transmission line is loaded with periodic SW components to create the SW transmission line. Let us analyze a single SW unit that is periodically cascaded over the transmission line. It has a period (p) and a susceptibility (b). The nth SW unit's voltage and current on both sides may be written as:

$$\begin{bmatrix} V_n \\ I_n \end{bmatrix} = \begin{bmatrix} A & B \\ C & D \end{bmatrix} \begin{bmatrix} V_{n+1} \\ I_{n+1} \end{bmatrix} \tag{2.9}$$

where A, B, C, and D are the matrix parameters of a cascade of a transmission line of length $p/2$, susceptance b, and another $p/2$ transmission line and can be expressed as:

$$\begin{bmatrix} A & B \\ C & D \end{bmatrix} = \begin{bmatrix} \cos\frac{kp}{2} & j\sin\frac{kp}{2} \\ j\sin\frac{kp}{2} & \cos\frac{kp}{2} \end{bmatrix} \begin{bmatrix} 1 & 0 \\ jb & 1 \end{bmatrix} \begin{bmatrix} \cos\frac{kp}{2} & j\sin\frac{kp}{2} \\ j\sin\frac{kp}{2} & \cos\frac{kp}{2} \end{bmatrix}$$
$$= \begin{bmatrix} \cos\theta - \frac{b}{2}\sin\theta & j\left(\sin\theta + \frac{b}{2}\sin\theta - \frac{b}{2}\right) \\ j\left(\sin\theta + \frac{b}{2}\sin\theta - \frac{b}{2}\right) & \cos\theta - \frac{b}{2}\sin\theta \end{bmatrix} \tag{2.10}$$

Without loss of generality, considering the $+z$ direction as the propagation direction and setting the reference phase point at $z = 0$, the voltage and current can be calculated as:

$$V(z) = V(0)e^{-\gamma z} \tag{2.11}$$

$$I(z) = I(0)e^{-\gamma z} \tag{2.12}$$

There is a propagation constant difference e^{-p} between the voltage and current of the nth unit and the voltage and current of the $n+1$th unit; the voltage and current of the $n+1$th unit can be expressed as:

$$V_{n+1} = V_n e^{-\gamma p} \tag{2.13}$$

$$I_{n+1} = I_n e^{-\gamma p} \tag{2.14}$$

Substituting Eqs. (2.13) and (2.14) into Eq. (2.9), it can be obtained that:

$$\begin{bmatrix} V_n \\ I_n \end{bmatrix} = \begin{bmatrix} A & B \\ C & D \end{bmatrix} \begin{bmatrix} V_{n+1} \\ I_{n+1} \end{bmatrix} = \begin{bmatrix} V_{n+1}e^{\gamma p} \\ I_{n+1}e^{\gamma p} \end{bmatrix} \tag{2.15}$$

$$\begin{bmatrix} A - e^{\gamma p} & B \\ C & D - e^{\gamma p} \end{bmatrix} \begin{bmatrix} V_{n+1} \\ I_{n+1} \end{bmatrix} = 0 \tag{2.16}$$

In order to obtain a nonzero solution, the matrix determinant to the left of Eq. (2.16) must be equal to 0, that is:

$$AD + e^{2\gamma p} - (A + D)e^{\gamma p} - BC = 0 \tag{2.17}$$

Since the network is reciprocal, $AD - BC = 1$, and Eq. (2.17) can be simplified as:

$$e^{-\gamma p} + e^{\gamma p} = A + D$$
$$\cosh\gamma p = \frac{A + D}{2} = \cos\theta - \frac{b}{2}\sin\theta \tag{2.18}$$

By substituting $\gamma = \alpha + j\beta$ into Eq. (2.18), it can be obtained that:

$$\cosh\gamma p = \cosh\alpha p\cos\beta p + j\sinh\alpha p\sin\beta p = \cos\theta - \frac{b}{2}\sin\theta \tag{2.19}$$

Given that the equation's right side (2.19) is a pure real number, in order for the equation to work, either α or β must be 0.

When $\alpha = 0$ and $\beta \neq 0$, the SW transmission line is capable of carrying EM waves. At this time, the SW transmission line has similar propagation properties to the traditional transmission line, except that the phase velocity and the guiding wavelength will be reduced, which can be utilized in the miniaturization of microwave components and antennas.

When $\alpha \neq 0$ and $\beta = 0$, it indicates that the EM wave cannot propagate on the SW transmission line. At this time, the SWS becomes the electromagnetic band gap (EBG) structure, which can be applied to antenna decoupling [1, 2], gap waveguide [3–6], etc.

Since the SW substrate-integrated waveguide (SIW) studied in this book is a waveguide-type transmission line, it is essential to analyze the field distribution. By loading the SWS in the traditional SIW transmission line, the electric or magnetic field can be concentrated into a certain region, akin to removing the original substrate in the region and replacing it with a substrate with high permittivity (electric field concentration) or high permeability (magnetic field concentration). Therefore, the equivalent relative permittivity (electric field concentration) [7] or permeability (magnetic field concentration) [8] of the entire SIW is increased, thereby lowering the antennas' and microwave components' physical dimensions while maintaining their electrical dimensions.

The phase velocity of an SW EM wave is less than the speed of light. The propagation characteristics of SW can also be obtained by solving Maxwell's equations:

$$\begin{aligned} \nabla \times \vec{E} &= -\frac{\partial \vec{H}}{\partial t} \\ \nabla \times \vec{H} &= \frac{\partial \vec{D}}{\partial t} + \vec{J} \\ \nabla \cdot \vec{D} &= \rho \\ \nabla \cdot \vec{B} &= 0 \end{aligned} \tag{2.20}$$

where $\vec{E}$, $\vec{H}$, $\vec{D}$, $\vec{B}$, $\vec{J}$, and ρ are the electric field intensity, magnetic field intensity, electric displacement vector, magnetic induction intensity, current density, and charge density, respectively, and $\vec{D}$ and $\vec{B}$ can be calculated as:

$$\begin{aligned} \vec{D} &= \varepsilon \vec{E} \\ \vec{B} &= \mu \vec{H} \end{aligned} \tag{2.21}$$

where ε is permittivity, and μ is permeability.

The phase constant and phase velocity can be calculated as:

$$\beta = \sqrt{\omega^2 \mu \varepsilon - k_c^{\ 2}} \tag{2.22}$$

$$v_p = \frac{\omega}{\beta} = \frac{1}{\sqrt{\mu\varepsilon - \frac{k_c^2}{\omega^2}}} \tag{2.23}$$

From Eqs. (2.22) and (2.23), it can be seen that the essence of SW miniaturization is to increase the equivalent relative permittivity and permeability of the transmission line, increase the phase shift constant, reduce the phase velocity, and shorten the waveguide wavelength under the premise of a constant operating frequency. Since the size of microwave components and antennas is wavelength-dependent, the physical size of microwave components and antennas can be miniaturized by loading SWSs without changing the electrical size.

SW propagation modes are usually characterized by SW coefficients. The intensity of the SW influence is indicated by the SW coefficient, which is the ratio of the light velocity to the phase velocity of the SW, and it can be calculated as:

$$SWF = \frac{c_0}{v_p} = \frac{\lambda_0}{\lambda_g} = \frac{\beta}{\beta_0} \tag{2.24}$$

2.2 Slow-Wave Substrate-Integrated Groove Gap Waveguide

2.2.1 Introduction

The substrate-integrated gap waveguide (SIGW) is a technology realized through printed circuit board (PCB) techniques. It has many benefits, including cheap cost, high integration level, and self-packaging. The applications of SIGW may be increased by reducing its size. Using SWSs is one efficient method of reducing the size of transmission cables. A number of SIWs with integrated SWSs have been developed recently. Prior studies have shown that applying wave effects to certain waveguide constructions may result in dimension reductions. For example, the cutoff frequency and phase velocity have been reduced by using an SW-SIW with loaded metalized via-holes. However, since multilayer PCB lamination technology is used, this method is expensive. To enhance the equivalent permeability and generate SW effects, another technique uses single-layer SW-SIWs designed with lumped inductors and microstrip polyline. However, both methods have a large transmission loss, particularly in high-frequency bands. Thus far, there has been no proposal for an SW substrate-integrated groove gap waveguide (SW-SIGGW).

This chapter presents a new method based on asymmetric half-height EBG structures, which we name SW-SIGGW. To provide a noticeable sluggish effect, this design embeds three rows of metalized via-holes in the waveguide. We may adjust the thickness of each layer while keeping the SIGGW's overall height constant by using asymmetric heighted mushroom EBG structures. In order to achieve a balance between preserving transmission performance, such as bandwidth and reflection coefficient, and reducing dimension via an SW effect, we optimize the height of via-holes. By lowering the phase velocity and waveguide cutoff frequency, an SW effect is applied, which decreases both longitudinal and transverse dimensions. We provide comprehensive details on our SW-SIGGW's design methodology and structure in Section 2.2.2. Furthermore, using measurements on a prototype, we describe how our results show a 41% decrease in lateral and longitudinal dimensions. Designing microwave components like couplers, filters, and power dividers with far lower dimensions than previous designs based on other technologies like multilayer PCB lamination or lumped inductor methods is made possible by the SIGGW.

2.2.2 Configuration of the Slow-Wave GW

2.2.2.1 Asymmetric Half-Height Mushroom Unit Cell

Figure 2.2a shows the unit cell for our suggested half-height mushroom-type EBG structure. It is made up of two mushroom-shaped structures that are separated by a dielectric layer. Each layer of Rogers RT5880 substrate material has $\varepsilon_r = 2$ and $\tan\delta = 0.0009$.

The bandgap of the symmetric half-height mushroom-type unit cell is less than that of the asymmetric half-height unit cell. Figure 2.3 shows the mushroom unit cell's dispersion curve with different t_1 and t_2 while keeping the overall thickness constant. The stopband clearly expands by 20% as the half-height mushroom unit cell gets more asymmetric, going from 9.14–21.4 GHz to 8.73–26.4 GHz. Asymmetric EBGs may provide an EM packaging structure in the operating band of mode 3 (30–46 GHz), where the EBG has no propagation mode, and the whole stopband between modes 1 and 2 (8.73–26.4 GHz), as shown in Figure 2.2b. The selection of a suitable height for the SW via-holes in Section 2.2.2.3 is made easier by the availability of substrates with varying thicknesses. Furthermore, even with a lower cutoff frequency, asymmetric half-height mushrooms have the benefit of

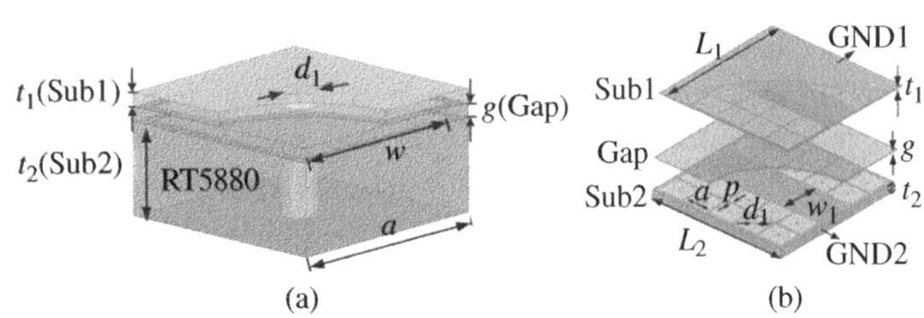

Figure 2.2 (a) Asymmetric half-height mushroom-type EBG structure. (b) Structure of SIGGW ($a = 2.3$ mm, $p = 0.3$ mm, $w = 2$ mm, $d_1 = 0.3$ mm, $t_1 = 0.127$ mm, $t_2 = 0.787$ mm, $g = 0.127$ mm, $W_1 = 3.56$ mm, $L_1 = 12.5$ mm, and $L_2 = 11.3$ mm).

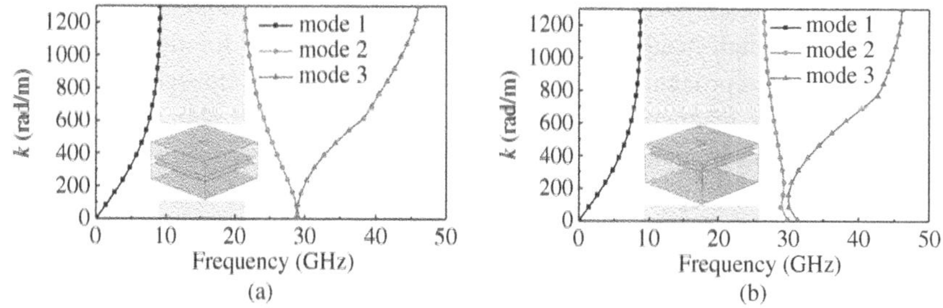

Figure 2.3 Dispersion diagram of the asymmetric half-height mushroom-type unit cell with different t_1 and t_2 [$t = t_1 + t_2 + g = 1.041$ mm]. (a) $t_1 = 0.457$, $t_2 = 0.457$, and $g = 0.127$. (b) $t_1 = 0.127$, $t_2 = 0.787$, and $g = 0.127$.

a broader stopband, which removes the need to rebuild the EBG structure when putting forth the SW-SIGGW in Section II-C.

2.2.2.2 Substrate-Integrated Groove GW

First, as shown in Figure 2.3b, a SIGGW with an asymmetric half-height mushroom EBG structure is built. The SIGGW's cutoff frequency is determined by its width, W_1. With the exception of the frequency range of mode 2 stimulated by the SIGGW, the simulated S-parameters of the SIGGW, shown in Figure 2.4, show good transmission performance above the cutoff frequency of 20 GHz.

2.2.2.3 Slow-Wave Substrate-Integrated Groove GW

The design and properties of an SWS implanted in the Sub2 layer that permits SW-SIGGW functioning are covered by the authors in this section. The height of the SWS, or "t_2," is closely correlated with the SW effect, according to the scientists. A high number for t_2 is preferred in order to optimize the SW effect. Impedance matching with feed lines becomes difficult, however, when periodic via-holes with a high t_2 value are used, since they may drastically change the waveguide's characteristic impedance. Consequently, the SW effect and the

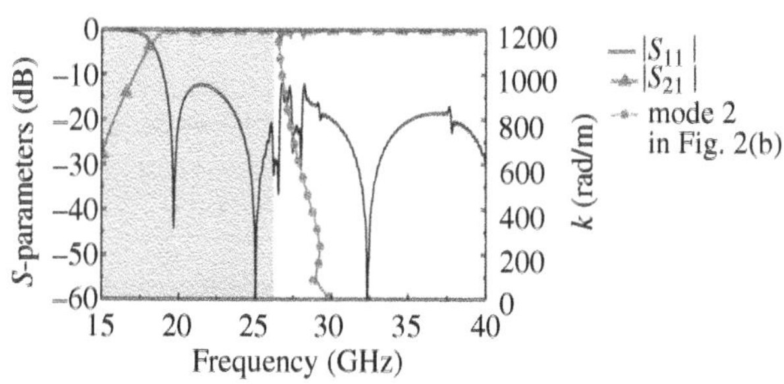

Figure 2.4 S-parameters of SIGGW.

SIGGW's transmission performance must be balanced. The authors suggest using an asymmetric half-height mushroom-type EBG structure for the SIGGW in order to solve this problem. The height of each layer may be adjusted more freely thanks to this arrangement. The authors back up their strategy with reference [7].

The authors clarify that the electric field strength is mostly focused in the top layer (Sub1) and the gap layer, with little effect on the distribution of the magnetic field, since the periodic metallic via-holes are loaded in the Sub2 layer. Based on references, they provide formulas for estimating the structure's size and SW impact [8].

$$\varepsilon_{reff} = \varepsilon_r \cdot \frac{t}{t_1 + g} = \varepsilon_r \cdot \left(1 + \frac{t_2}{t_1 + g}\right) \tag{2.25}$$

$$f_{csw} = \frac{c_0}{2 \cdot W_{eff} \cdot \sqrt{\varepsilon_{reff}}} = \sqrt{\frac{\varepsilon_r}{\varepsilon_{reff}}} f_c \tag{2.26}$$

$$v_\varphi = \frac{c_0}{\sqrt{\varepsilon_{reff} \cdot \left(1 - \left(\frac{f_{csu}}{f}\right)^2\right)}} \tag{2.27}$$

Slow-wave factor (SWF) is defined as v_ϕ / c_0

$$SWF = v_\varphi / c_0 = \frac{1}{\sqrt{\varepsilon_{reff} \cdot \left(1 - \left(\frac{f_{csu}}{f}\right)^2\right)}} = \frac{1}{\varepsilon_r \cdot \frac{t}{t_1+g} - \varepsilon_r \cdot \frac{f_c^2}{f^2}} \tag{2.28}$$

The SWS will cause the cutoff frequency of SW-SIGGW to drop, as inferred from (2.1) and (2.2). In comparison to standard SIGGW, the electrical transverse dimension of SW-SIGGW normalized by free-space wavelength will be less. Figure 2.5b displays the SWF versus frequency for SIGGW with varying SWS

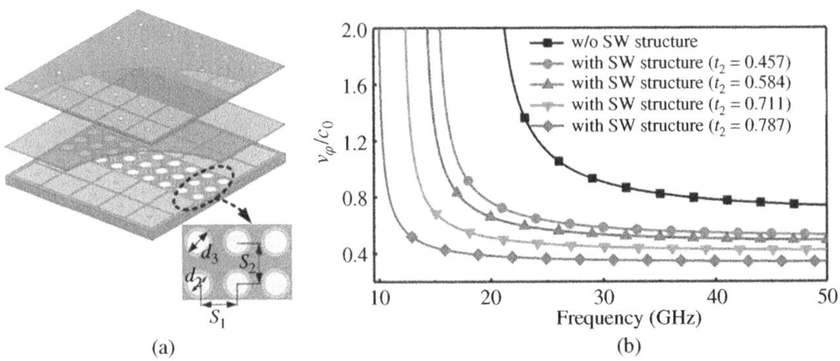

Figure 2.5 (a) Structure of SW-SIGGW ($S_1 = 1.2$ mm, $S_2 = 1.25$ mm, $d_2 = 0.7$ mm, and $d_3 = 0.92$ mm). (b) SWF of SIGGW with/without slow-wave structure.

height. It demonstrates how the SW effect increases with via-hole height. According to (2.4), loading the SWS with $t_2 = 0.787$ mm results in a 56% drop in phase velocity when compared to SWF without the SWS.

2.2.2.4 Transition to Microstrip Line

For the Ku-band, a transition from a 50 Ω microstrip line to SW-SIGGW is planned, as shown in Figure 2.6. Figure 2.7 shows the transition's design process. The SW-SIGGW with the first transition, which includes a microstrip line and a via-hole transition, is shown in Figure 2.7a. The via-hole transition connects the microstrip line to the SW-SIGGW's bottom layer and functions as a 3-D signal routing geometry. The microstrip's width between points A and B is 0.83 mm.

The first transition's impedance matching on the Ku-band is insufficient, as shown in Figure 2.8. As a result, the switch from the microstrip feed line to the GW

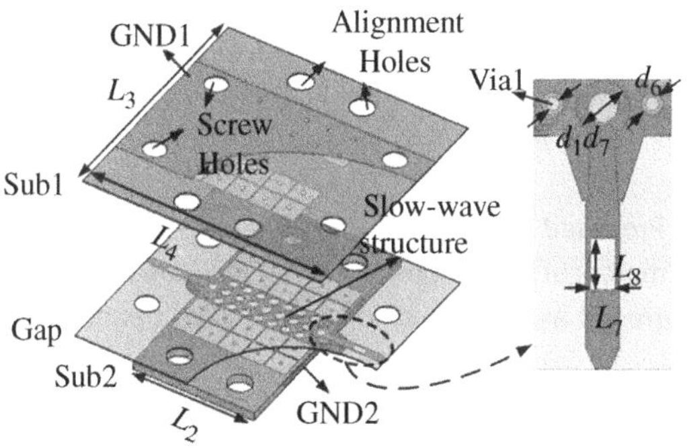

Figure 2.6 Structure of SW-SIGGW with microstrip transition ($L_3 = 22.56$ mm, $L_4 = 22.1$ mm, $d_6 = 0.52$ mm, $d_7 = 1.28$ mm, $L_7 = 0.6$ mm, and $L_8 = 1.2$ mm).

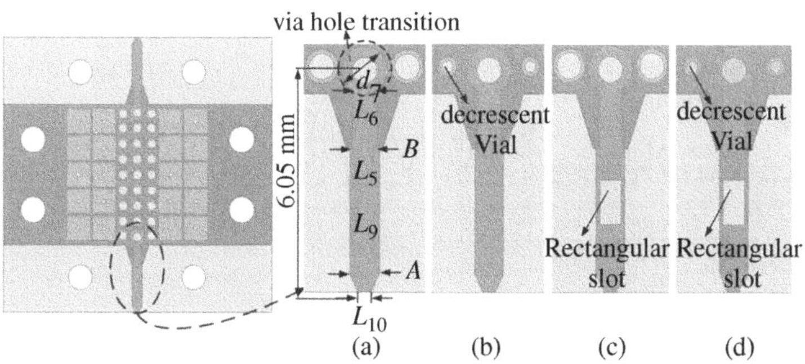

Figure 2.7 Evolution process of the transition ($L_5 = 0.62$ mm, $L_6 = 0.69$ mm, and $L_9 = 0.83$ mm). (a) Initial transition. (b) Transition with only decrescent Via1. (c) Transition with only a rectangular slot. (d) Improved transition.

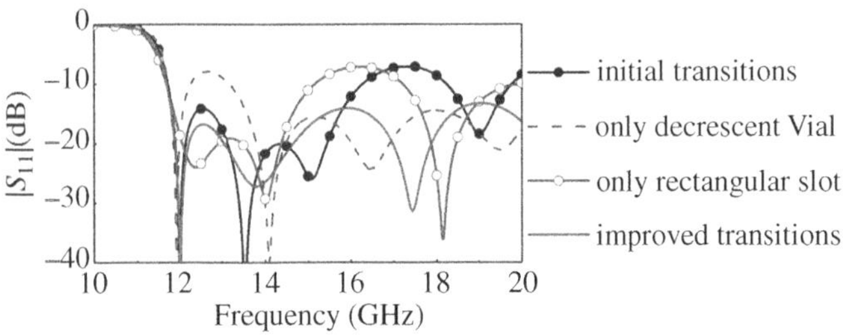

Figure 2.8 $|S_{11}|$ of the SW-SIGGW for each stage of the evolution process of the transition.

is altered. S_{11} is enhanced across 15.81–18.62 GHz by decrescent Via1 on both sides of the feed line, as shown in Figure 2.7b. Furthermore, as shown in Figure 2.7c, a rectangular slot etched on the feed line enhances S_{11} across 11.87–14.64 GHz. The effective implementation of the improved transition is shown in Figure 2.8, which shows the simulated S-parameters of the SW-SIGGW transition at each step of the evolution process, attaining a reflection coefficient greater than −14 dB over the whole Ku-band.

2.2.2.5 Slow-Wave Effect of the Proposed SW-SIGGW

The field distribution and S-parameters of the proposed SW-SIGGW are compared with those of the SIGGW in Figure 2.9 in order to evaluate the SW impact of the SW-SIGGW.

It is clear from Figure 2.9a that L_2 = 11.3 mm denotes one-and-a-half guided wavelengths for SIGGW at 30 GHz and SW-SIGGW at 17.7 GHz. This shows how the SWS successfully reaches 41%.

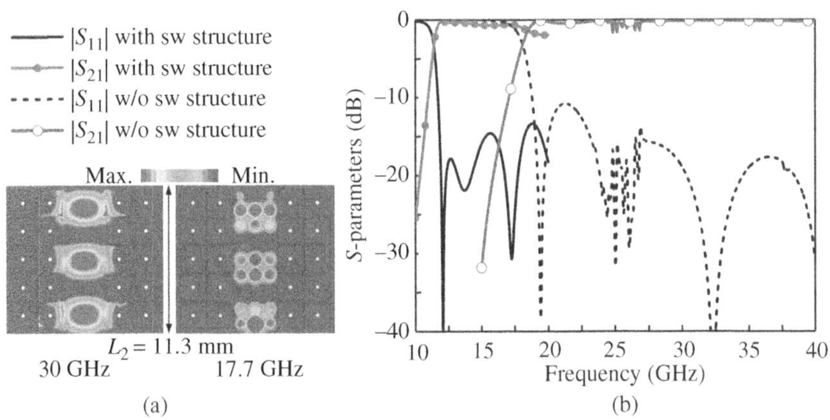

Figure 2.9 (a) E-field distribution. (b) S-parameters with/without SW structure.

2.2.3 Manufacture and Measurement

Using PCB methods, an SW-SIGGW prototype is created and measured in a microwave chamber. The microstrip lines are printed on the gap layer's bottom layer and Sub2's top layer for manufacturing viability. The mounting bed is a copper plate that is 0.5 mm thick. Figure 2.10 shows pictures of the constructed circuits, and the same figure shows the S-parameters of the suggested SW-SIGGW for simulations and measurements that were acquired using the commercial full-wave solver CST Microwave Studio.

From 12 to 18 GHz, the observed reflection coefficient is better than −10 dB across the Ku-band. Taking into account two transitions, the maximum insertion loss at 18 GHz is 0.93 dB. The main causes of the differences between the observed and simulated scattering coefficients include assembly mistakes, soldering faults in the connectors, and manufacturing tolerances.

Table 2.1 compares a number of important SW-SIGGW properties with those of the SW-SIW, including cutoff frequency, insertion loss, and transverse electrical dimension at the cutoff frequency. Despite operating at a much higher frequency, a comparison shows that our SW-SIGGW outperforms the SW-SIW in [9, 10] and shows equivalent insertion loss to the SW-SIW in [8].

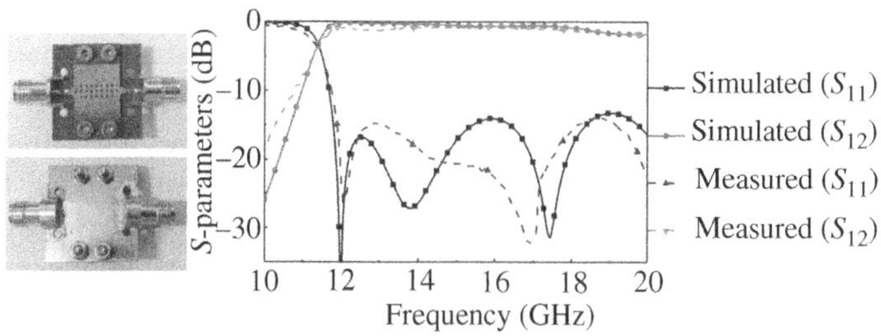

Figure 2.10 Simulated and measured S-parameters of SW-SIGGW.

Table 2.1 Comparison of the Proposed SW-SIGGW and the SW-SIW.

Reference	Type	Cutoff frequency (GHz)	IL(dB)	Transversal electrical dimension at the cutoff frequency
[8]	SW-SIW	9.3	0.63	$0.327\lambda_g$
[9]	SW-SIW	5.2	0.75	$0.357\lambda_g$
[10]	SW-SIW	2.95	1.5	$0.328\lambda_g$
This work	SW-SIGGW	12	0.7	$0.741\lambda_g$

2.2.4 Conclusion

An SW-SIGGW is introduced for the first time in this chapter. A detailed demonstration of the SW mechanism and the design process is given. The cutoff frequency and phase velocity are reduced by 41% as compared to the conventional groove GW with the same lateral and longitudinal dimensions. To validate the suggested design, an SW-SIGGW prototype is successfully created, built, and measured. Designing microwave components with substantially smaller dimensions, such as couplers, filters, and power dividers, is possible with the greatly reduced SW-SIGGW. In addition, these SW-SIGGW-based components will be less expensive and easier to fabricate than the SW-SIW-based components in [8].

2.3 Slow-Wave Substrate-Integrated Waveguide with Miniaturized Dimensions and Broadened Bandwidth

2.3.1 Introduction

Due to the rapid advancement of millimeter-wave and microwave technology, circuits are becoming more compact, low-profile, lightweight, and inexpensive. A potential planar transmission line that makes use of PCBs and metallic rectangular waveguides is called a substrate-integrated waveguide (SIW). It benefits from low loss, lightweight design, planar integration, and reduced expenses. Its primary disadvantage, meanwhile, is its wide lateral breadth, which makes it challenging to use in small circuits. Consequently, an SWS was used to create an SW propagation mode that lowers the EM wave's phase velocity. Longitudinal transmission lines and circuit miniaturizations may be achieved by decreasing the guided wavelength at a certain frequency. This is accomplished by current SW-SIWs by improving effective permeability or permittivity. Nevertheless, along the periodic SWS, this method leads to either a higher wave impedance for inductive loading or a lower wave impedance for capacitive loading. This results in a restricted operating bandwidth while retaining a high SW factor (SWF) due to poor impedance matching between SWS-loaded parts and unloaded regions of SW-SIW.

In this chapter, by concurrently raising effective permeability and permittivity, we suggest an SW-SIW with improved SWF and expanded bandwidth. This is accomplished by shorting patches to metalized blind via-holes and loading them, which simultaneously improves effective permittivity and permeability. Our suggested design has smaller longitudinal dimensions and smaller lateral dimensions because of greater SWF values and lower cutoff frequencies, respectively, in contrast to previously published SW-SIW structures that just enhance effective permeability or permittivity. Additionally, stable wave impedances across the periodic SWS allow our suggested design to attain broader bandwidths.

In comparison to typical SIW structures with comparable cutoff frequencies, the lateral dimension has been lowered by 53% while keeping the same performance attributes. Additionally, we were able to reduce the phase velocity by 73%, which led to reduced longitudinal dimensions for certain electrical lengths. With a fractional bandwidth coverage of around 91%, the operating band we were able to attain spans 7–18 GHz. A prototype built using our suggested design has been tested, and the measurement results show good agreement with the models.

2.3.2 SW-SIW Operation Principle

2.3.2.1 Configuration

Figure 2.11 shows the suggested SW-SIW's transversal cross section and 3-D schematic view. Two Rogers 4350B substrates with a dielectric loss tangent $\tan\delta = 0.0037$ and a relative dielectric constant $\varepsilon_r = 3.66$ are used to build the whole periodic structure. The width of the SW-SIW, W_1, is set at 5.7 mm, and the thickness of the top and bottom substrates is $h_1 = 0.254$ mm and $h_2 = 0.762$ mm,

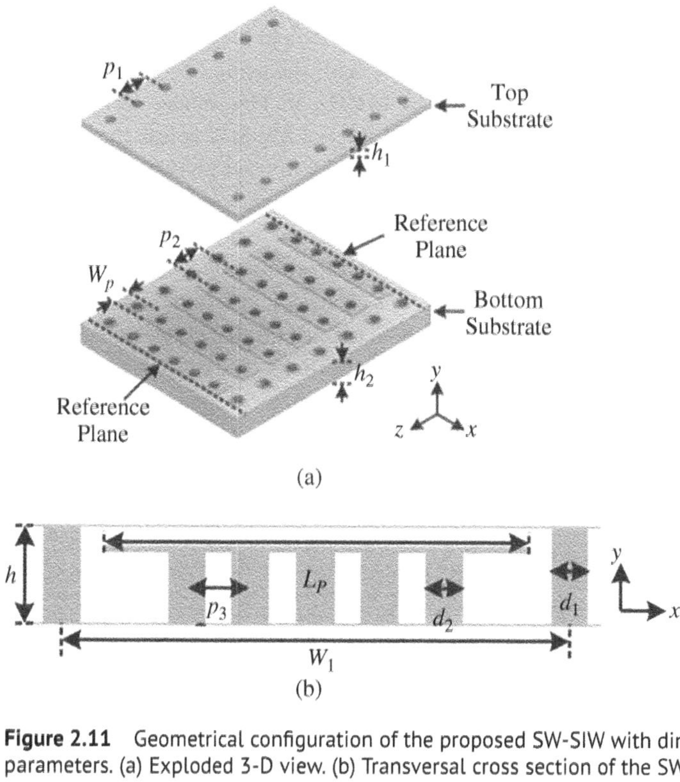

Figure 2.11 Geometrical configuration of the proposed SW-SIW with dimensional parameters. (a) Exploded 3-D view. (b) Transversal cross section of the SW-SIW.

respectively. In accordance with SIW design requirements, the via-holes' period (p_1) and diameter (d_1) are set at 0.8 and 0.4 mm, respectively [11].

Five rows of metalized blind via-holes with a diameter of d_2 and a separation distance of p_3 are integrated into the bottom substrate. The top surface of the bottom substrate is printed with rectangular metal patches that are electrically coupled to the blind via-holes and have length L_p and width W_p. At a time p_2, the metalized blind via-holes and patches are sporadically spaced along the longitudinal direction.

2.3.2.2 Slow-Wave Effect

The rectangular patch shorted to the bottom plane of the SIW via blind via-holes and metalized blind via-holes makes up the proposed SWS unit. The effective permeability is increased by stretching the current channel by placing patches on the blind via-holes, while the effective permittivity is increased by the extra capacitance created between the patch and the top ground plane. When such SWSs are loaded, both their effective permittivity and permeability rise at the same time, raising the SWF.

Ansys HFSS is used to do complete wave simulations of the suggested structure in order to clarify the SWS process. The electric field magnitude distributions on the H-plane in SW-SIW and normal SIW with the same lateral and longitudinal dimensions at a cutoff frequency of 15 GHz, in TE_{10} mode, are shown in Figure 2.12. Notably, the guided wavelength (λ_g) of SW-SIW is less than that of standard SIW at this frequency, suggesting that the EM wave phase velocity inside SW-SIW is lower.

Figures 2.13 and 2.14 show the electric and magnetic field magnitude distributions on the cross section of the regular SIW and the proposed SW-SIW. It is clear from Figure 2.13 that the conventional SIW's electric field is homogeneous along the y-axis, whereas the SW-SIW's electric field is concentrated in the top substrate, as shown in [8], leading to a higher effective permittivity. The suggested SW-SIW's magnetic field is mostly centered in the bottom substrate, as shown in Figure 2.15.

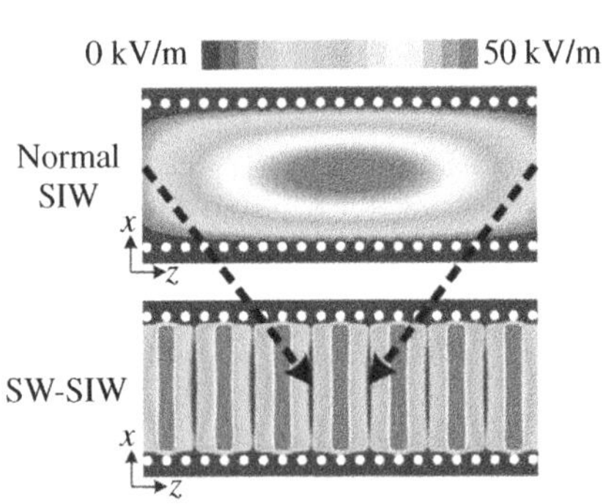

Figure 2.12 Wavelength reduction of SW-SIW compared with normal SIW at cutoff frequency 15 GHz.

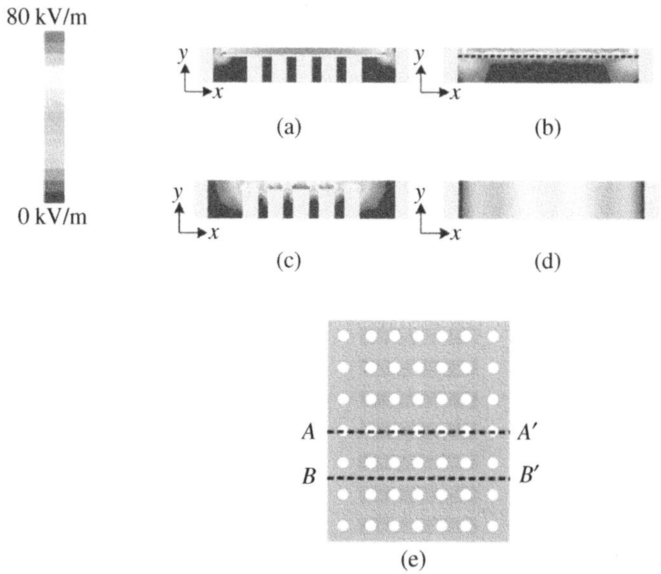

Figure 2.13 *E*-field magnitude distribution on different cross-sectional positions. (a) *E*-field on *AA′* cross section with patch. (b) *E*-field on *BB′* cross section with patch. (c) *E*-field on *AA′* cross section without patch. (d) *E*-field in normal SIW. (e) Cross-sectional position.

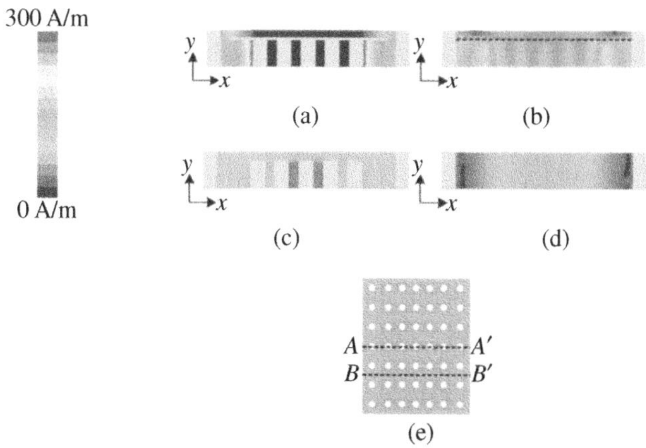

Figure 2.14 *H*-field magnitude distribution on different cross-sectional positions. (a) *H*-field on *AA′* cross section with patch. (b) *H*-field on *BB′* cross section with patch. (c) *H*-field on *AA′* cross section without patch. (d) *H*-field in normal SIW. (e) Cross-sectional position.

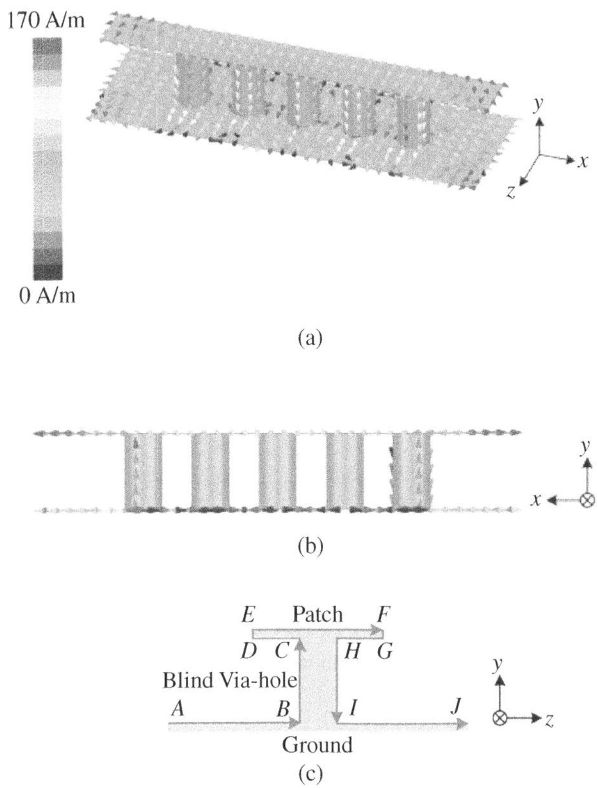

Figure 2.15 Current vector distribution on the SWS in a unit element of the SW-SIW. (a) 3-D view. (b) Back view. (c) Heuristic diagram.

A little, but weaker, component is also present at the top substrate's border. About three times as much magnetic field is present in the bottom substrate as there is at the patch border in the top substrate. According to [9], this magnetic field concentration is a sign of increased effective permeability. SW transmission lines, which spatially separate electric and magnetic fields, are characterized by this distribution of EM fields [12].

Figure 2.15 shows the distribution of the current vector on the SWS in a unit element of the SW-SIW. After passing through one side of the blind via-hole *B–C* and the top and lower surfaces of the shorted patch *C–D–E–F–G–H*, the current returns to the ground plane *I–J* and the opposite side of the blind via-hole *H–I*. The suggested SW-SIW's current path is comparable to that of a meandering strip-type inductor with pins linked at locations *B* and *I*. Along the path of propagation, the equivalent inductance rises. Since the inductor is known to be able to capture the magnetic field, the bottom substrate of the proposed SW-SIW is where the magnetic field is focused, as shown in Figure 4.13. As shown in [9], increasing the effective permeability often results in an increase in the magnetic concentration. Additionally, loading patches on the blind via-holes lengthens the current route compared to a single blind via-hole without the patch. An increase in the inductance along the transmission line's propagation direction corresponds to an increase in the current path along that direction, which raises the effective permeability.

As the bottom substrate's magnetic field is restricted and the top substrate's electric field is confined, the effective relative permittivity ε_{r_eff} can be preliminarily obtained by the following equations [8]:

$$\varepsilon_{r\_eff} = \varepsilon_r \cdot \frac{h}{h_1} = \varepsilon_r \cdot \left(1 + \frac{h_2}{h_1}\right) \tag{2.29}$$

Similarly, the permeability μ_{r_eff} can be obtained by

$$\mu_{r\_eff} = \mu_r \cdot \frac{h}{h_1} = \mu_r \cdot \left(1 + \frac{h_2}{h_1}\right) \tag{2.30}$$

where ε_r and μ_r are the relative permittivity and permeability of the substrate, respectively. The cutoff frequency of the SW-SIW for TE_{10} mode can be derived as

$$f_{c\_SW-SIW} = \frac{c_0}{2 \cdot W_{eff} \cdot \sqrt{\mu_{r\_eff} \cdot \varepsilon_{r\_eff}}} \tag{2.31}$$

where c_0 is the light velocity in free space, and W_{eff} is the effective width of the SIW and can be obtained by [13]:

$$W_{eff} = W_1 - 1.08 \cdot \frac{d_1^2}{p_1} + 0.1 \cdot \frac{d_1^2}{W_1}. \tag{2.32}$$

The phase velocity of the SW-SIW can be derived by the following equation:

$$v_p = \frac{c_0}{\sqrt{\mu_{r\_eff} \cdot \varepsilon_{r\_eff} \cdot \left(1 - \left(\frac{f_{c\_SW-SIW}}{f}\right)^2\right)}} \tag{2.33}$$

The guided wavelength can be derived as

$$\lambda_g = v_p / f \tag{2.34}$$

Equations (2.5) through (2.10) show that as the effective permittivity and permeability rise, the loading of the SWS decreases the cutoff frequency, phase velocity, and directed wavelength. The normalized phase velocity computed using Eq. (2.9) is compared to the value simulated by CST, as shown in Figure 2.16, in order to verify the correctness of the calculation model produced from these equations. It should be mentioned that Eqs. (2.5)–(2.10) do not take into consideration other size characteristics that might affect the phase velocity, despite the fact that the error between the two computations is minimal.

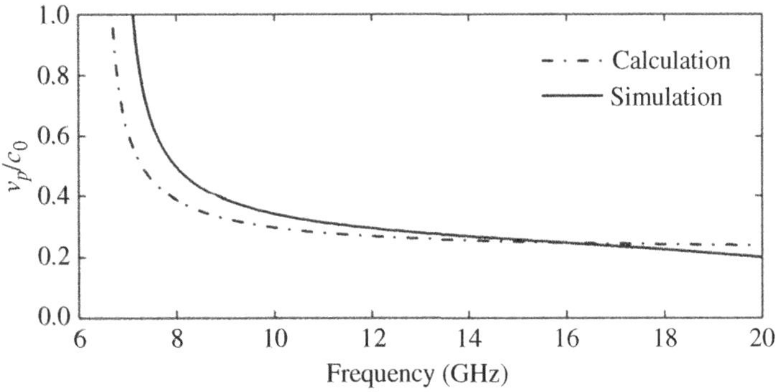

Figure 2.16 Calculated and simulated normalized phase velocity of SW-SIW ($h_1 = 0.254$ mm and $h_2 = 0.762$ mm).

2.3.2.3 Parametric Study

The aforementioned study indicates that the effective permittivity depends on the blind via-hole height in relation to the SW-SIW's overall thickness, h_2/h, the loaded patches' length and breadth, L_p and W_p, and the lateral distance between the blind via-holes, p_3. Conversely, the number and diameter of the blind via-holes, n and d_2, and the patch width, W_p, have the most effects on the effective permeability. We carried out several parameter experiments to graphically illustrate how these factors affected the SW. We used eigenmode analysis in CST to simulate a unit element of the SW-SIW with periodic bounds in the propagation direction in order to determine the phase velocity. The size parameters of the suggested SWS are set to the following values for our parameter studies: $h_2 = 0.762$ mm, $L_p = 4.6$ mm, $W_p = 0.6$ mm, $p_3 = 0.7$ mm, and $d_2 = 0.4$ mm.

Figure 2.17 shows how the height h_2 of the blind via-holes affects the cutoff frequency and the normalized phase velocity ($v\phi/c$) while keeping the overall substrate thickness h constant. The conventional SIW has a cutoff frequency of 15 GHz. The cutoff frequency and phase velocity both drop with increasing blind via-hole height h_2, suggesting a larger SW impact. Since capacitance is inversely proportional to the distance between the metal planes, this is explained by the increase in capacitance between the grounded patches and the top ground plane. The capacitance between the grounded patch and the upper ground plane is affected by the patch's length and breadth. Figures 2.18 and 2.19 show how the cutoff frequency and normalized phase velocity of SW-SIW are affected by patch width and length. Normalized phase velocities and reduced cutoff frequencies are the outcomes of increasing the patch's width (W_p) and length (L_p). The capacitance effect between the patches and the top ground plane is strengthened as L_p and W_p are bigger because the patch area grows. Furthermore, the current flowing down the grounded patch's short edge introduces an inductance

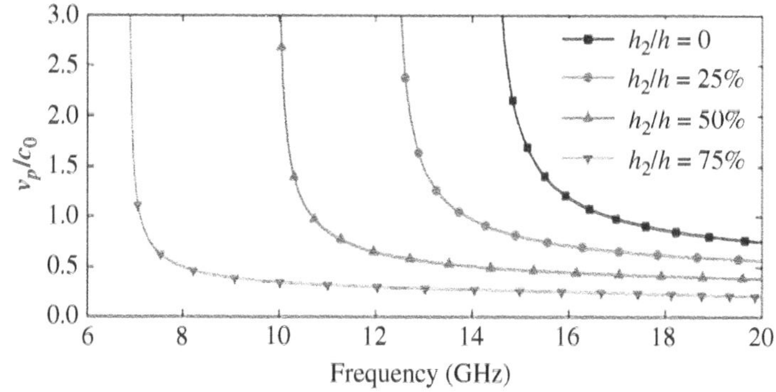

Figure 2.17 Simulated normalized phase velocity versus frequency for different values of the height of blind metalized via-holes ($h = 1.016$ mm).

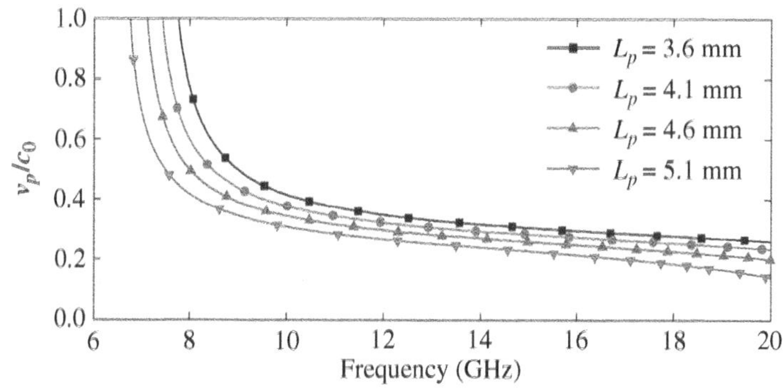

Figure 2.18 Simulated normalized phase velocity versus frequency for different values of the patch length.

component, which lowers the cutoff frequency and phase velocity even more as the width, W_p, increases. Figure 2.20 illustrates how the cutoff frequency and normalized phase velocity of SW-SIW are impacted by variations in the number of blind via-holes. Phase velocities and cutoff frequencies fall as the number of blind via-holes decreases. The equivalent shunt inductors in a unit cell of SW-SIW are determined by the number of blind via-holes. A unit cell's overall inductance rises as the number of blind via-holes falls, increasing the effective permeability and decreasing the phase velocity and cutoff frequency.

As shown in Figure 2.21, reducing the diameter of blind via-holes results in normalized phase velocities and lower cutoff frequencies. By lengthening the patch's current route, this decrease raises the effective permeability. On the other hand, the present route stays the same when the patch is not loaded on the blind

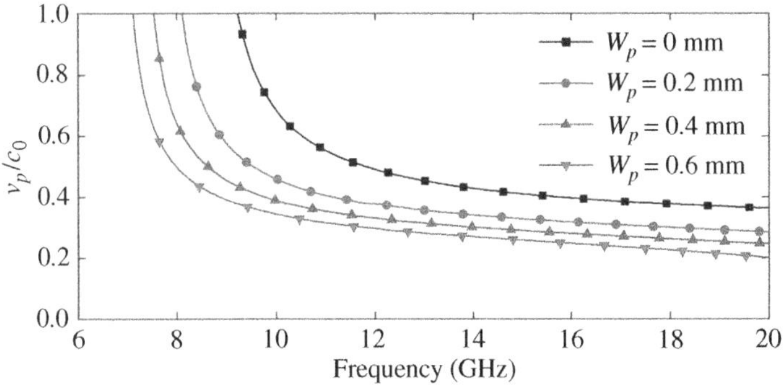

Figure 2.19 Simulated normalized phase velocity versus frequency for different values of the patch width.

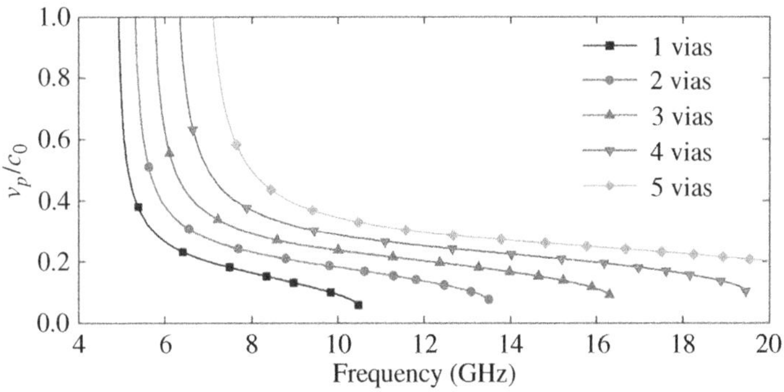

Figure 2.20 Simulated normalized phase velocity versus frequency for different values of the via-holes' number.

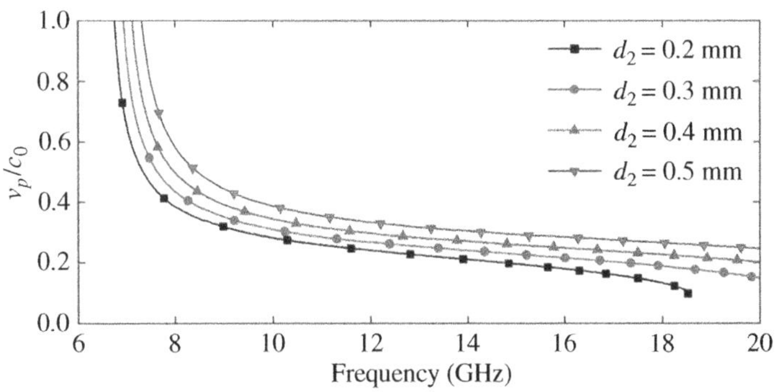

Figure 2.21 Simulated normalized phase velocity versus frequency for different values of the via-holes' diameter.

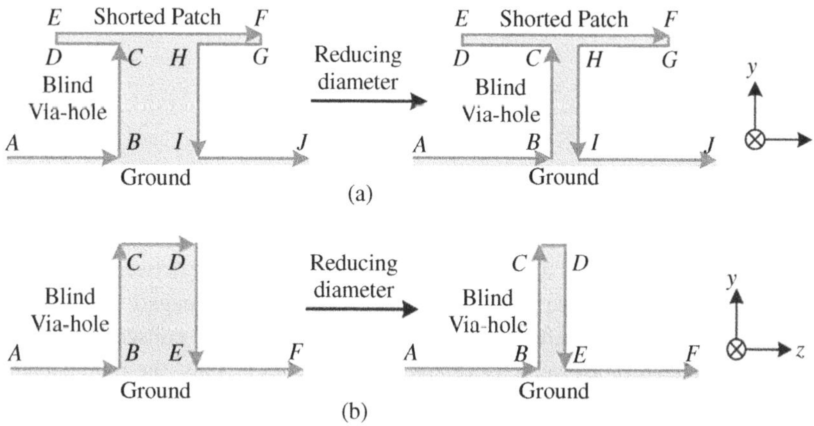

Figure 2.22 Heuristic diagram of the current path. (a) With patch. (b) Without patch.

via-hole (see Fig. 2.22). Additionally, lowering the blind via-hole diameter raises the via-hole's inductance, which raises SW-SIW's equivalent inductance and effective permeability [14].

$$L_{via-hole} = \frac{\mu_0}{2\pi} \cdot 4 \left| \ln\left(\frac{4h_2}{d_2}\right) - 1 \right| \cdot h_2 \tag{2.35}$$

Reducing p_3 results in a drop in both the cutoff frequency and the normalized phase velocity, as shown in Figure 2.23. The rise in reciprocal inductance between two nearby blind via-holes explains this occurrence. Two blind via-holes' mutual

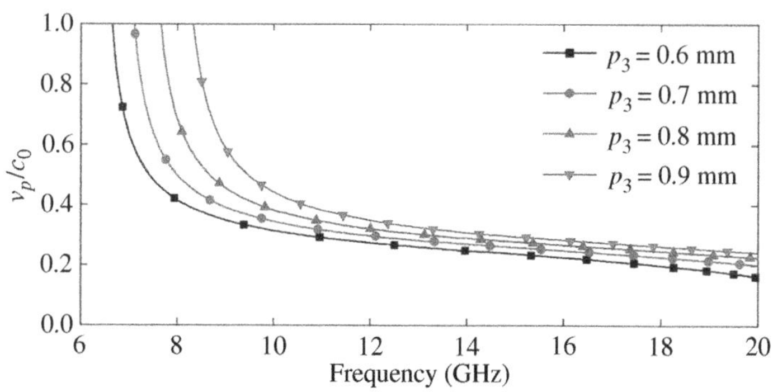

Figure 2.23 Simulated normalized phase velocity versus frequency for different values of lateral distance between two adjacent blind via-holes.

inductance may be determined by [14].

$$L_m = \frac{\mu_0}{2\pi}\left(\ln\left(\frac{h_2+\sqrt{h_2^2+p_3^2}}{p_3}\right)h_2+p_3-\sqrt{h_2^2+p_3^2}\right) \tag{2.36}$$

Effective permeability and equivalent inductance rise as a result of the mutual inductance increasing as p_3 decreases.

Furthermore, while it is usually insignificant, a capacitance between consecutive SWSs may affect the equivalent inductance along the propagation direction. Figure 2.24 illustrates the effect of the distance p_2 between two consecutive SWSs on the normalized phase velocity versus frequency. As p_2 falls, the total capacitance between two SWSs rises, but the phase velocity is little affected.

Lastly, Table 2.2 presents the final dimensions of the proposed SW-SIW based on the analysis provided. The SW-SIW may be optimized to reach the required degree of downsizing while increasing the operating bandwidth by adhering to the described design process.

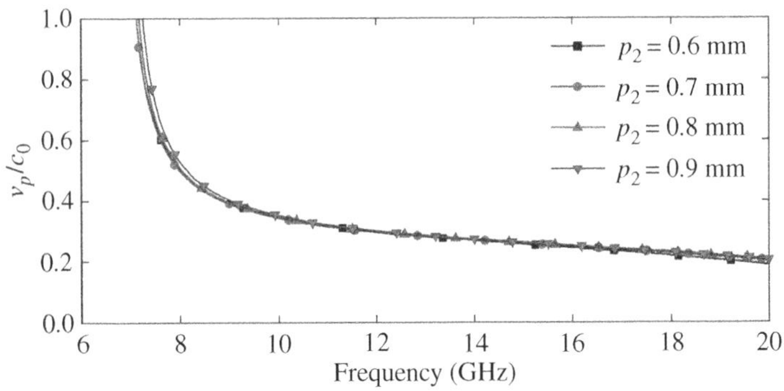

Figure 2.24 Normalized phase velocity versus frequency for different values of the distance between two successive SWSs p_2.

Table 2.2 Geometrical Parameters of SW-SIW (Units: Millimeters).

Variable	L_p	W_p	W_1	h_1
Value	4.6	0.6	5.7	0.254
Variable	h_1	h	d_1	d_2
Value	0.762	1.016	0.4	0.4
Variable	p_1	p_2	p_3	
Value	0.8	0.8	0.7	

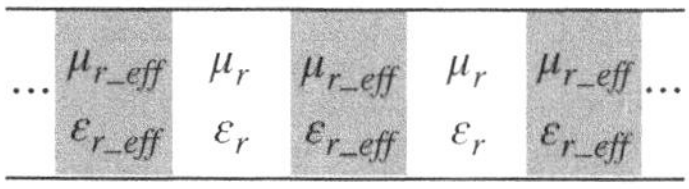

Figure 2.25 Heuristic diagram of the transmission line periodically loaded by SWSs.

2.3.2.4 Impedance Matching of SW-SIW

Figure 2.25 shows the heuristic diagram of the suggested SW-SIW with respect to medium characteristic parameters, showing that the SW-SIW is composed of an alternating SWS-loaded portion and an SWS-unloaded part. The following formula can be used to calculate the wave impedance of the SWS-loaded component for TE_{10} mode:

$$Z_w = \frac{\sqrt{\mu_{r\_eff}/\varepsilon_{r\_eff}}}{\sqrt{1-(\lambda/2a)^2}}. \tag{2.37}$$

There is a negative correlation between the wave impedance and the effective permittivity and a positive correlation with the effective permeability. As a result, either by raising effective permeability or by increasing effective permittivity, the wave impedance of the SWS-loaded portion of the SW-SIW will rise. Since the SWS may greatly lower the directed wavelength, its size is not necessarily substantially smaller than the guided wavelength. The electrical size of SWS increases, especially as the operating frequency rises, and the guided wavelength approaches the physical size of SWS. Poor impedance matching between SWS-loaded and SWS-unloaded portions of the SW-SIW is caused by the considerable variations in wave impedance along the periodic SWS. A mere increase in effective permeability or permittivity will often result in a narrowing of the SW-SIW bandwidth. In contrast to the SW-SIW with simply rising effective permeability or permittivity, the wave impedance of the suggested SW-SIW along the periodic SWS may be much more stable when both effective permittivity and permeability are raised concurrently; see (2.13). It is possible to greatly enhance the impedance matching between the SWS-loaded and SWS-unloaded components of SW-SIW, which will increase the operating bandwidth.

Figure 2.26 shows the *S*-parameters for the regular SIW and the suggested SW-SIW. Between 7.0 and 20.9 GHz is the frequency range where S_{11} is less than or equal to −10 dB. In contrast to the standard SIW, which has a cutoff frequency of 15 GHz, the SW-SIW has a cutoff frequency of 7.0 GHz, which is 53% lower.

2.3.3 Simulated and Measured Results

The final SW-SIW with the transition from SW-SIW to grounded coplanar waveguide (GCPW) configuration is shown in Figure 2.27. To demonstrate the viability of the proposed SW-SIW, an SW-SIW with the transition from SW-SIW to GCPW

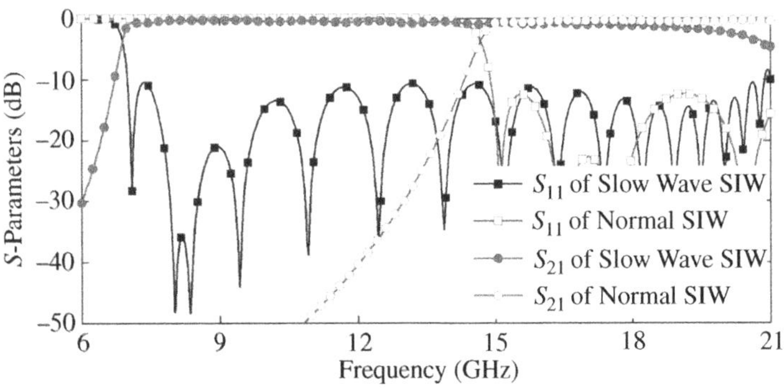

Figure 2.26 Simulated S-parameters of normal SIW and SW-SIW.

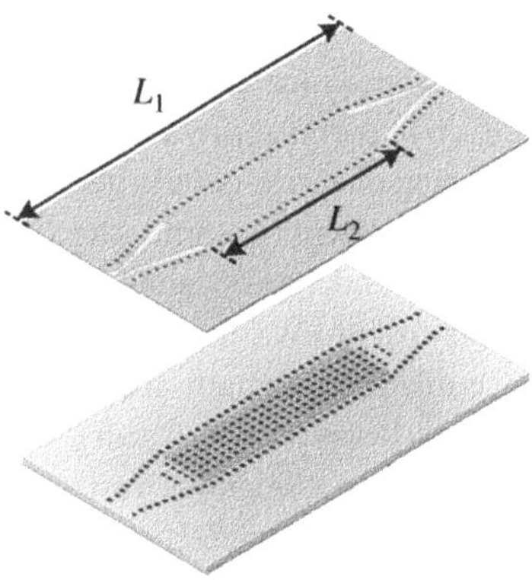

Figure 2.27 Exploded 3-D view of the final SW-SIW (L_1 = 35 mm and L_2 = 19.2 mm).

is planned, simulated, optimized, and manufactured based on the design rule in Section 2.3.2. Figure 2.28 displays the geometry of the GCPW–SW-SIW transition. As shown in Figure 2.29, a prototype of the SW-SIW with back-to-back transition is created using the PCB fabrication method on two Rogers 4350B substrates.

2.3.3.1 Transition From SW-SIW to GCPW

For the measurement feasibility, a switch froma SW-SIW to GCPW is created using the design process in [15]. Figure 2.28 displays the geometry of the GCPW–SW-SIW transition. The transition is made up of two tapered slots, two rows of metalized blind via-holes, a 50 Ω GCPW, and a tapered GCPW that is coupled to a coaxial connection. To prevent EM leakage and parallel-plate mode, two rows of metalized via-holes are inserted on each side of the GCPW-to-SW-SIW transition [15]. The tapered slot acts as an impedance transformer between GCPW and SW-SIW in addition to transmitting the EM wave from GCPW into SW-SIW.

In order to match impedance, the blind via-holes are additionally loaded within the transition. Additionally, in order to modify the wave mode in the SW-SIW,

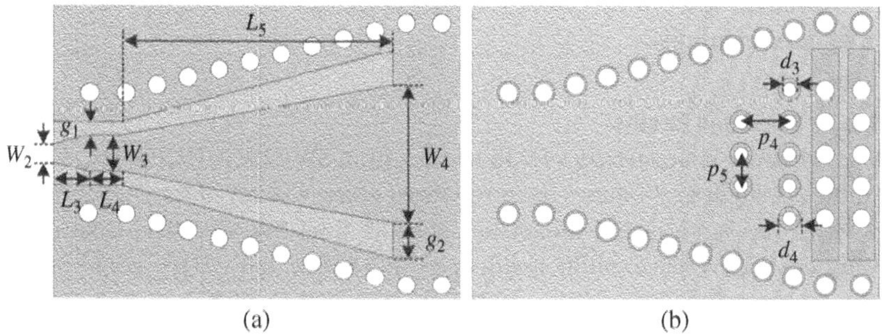

Figure 2.28 Configuration of the transition from SW-SIW to GCPW. (a) Top layer of the top substrate. (b) Top layer of the bottom substrate.

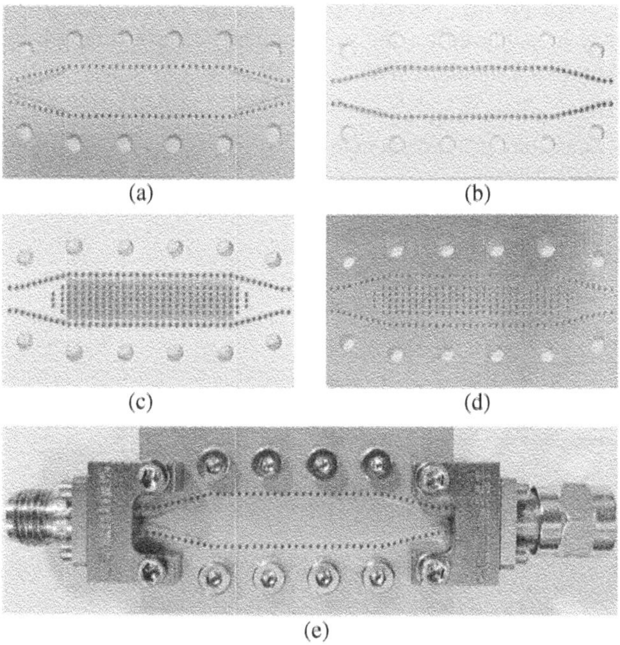

Figure 2.29 Photographs of the proposed SW-SIW with back-to-back transition. (a) Top layer of the top substrate. (b) Bottom layer of the top substrate. (c) Top layer of the bottom substrate. (d) Bottom layer of the bottom substrate. (e) Assembled prototype with *K*-connectors.

the blind via-holes may focus the electric field onto the upper substrate. To provide a progressive conversion of wave impedance, the number of blind via-holes per row is raised from three to five along the route from transition to SW-SIW. The capacitive effect increases and the wave impedance decreases as the blind via-hole diameter increases. The inductive effect intensifies and the wave impedance rises as the distance between two blind via-holes decreases. Table 2.3 lists the transition's final dimensions.

2.3.3.2 Simulated Results

The geometrical characteristics for the switch from SW-SIW to GCPW are shown in Table 2.3. Figure 2.30 shows the normalized phase velocity of the conventional SIW with the same lateral dimension and the final SW-SIW. The traditional SIW and SW-SIW have cutoff frequencies of 15 GHz and 7 GHz, respectively. When compared to the traditional SIW, the SW-SIW's cutoff frequency is 53% lower. Therefore, an SW-SIW may be used to generate a narrower waveguide with a 53% decrease in lateral dimension for the same required cutoff frequency. The SW-SIW's phase velocity is 73% lower than that of the traditional SIW, as shown in Figure 2.30. Consequently, a shorter waveguide with a 73% decrease in longitudinal length may be achieved for the same targeted electrical length by using an SW-SIW rather than a standard SIW.

Table 2.3 Geometrical Parameters of Transition from SW-SIW to GCPW (Units: Millimeters).

Variable	L_3	L_4	L_5	W_2
Value	0.8	0.7	6.1	0.4
Variable	W_3	W_4	d_3	d_4
Value	0.6	4.5	0.3	0.5
Variable	p_4	p_5	g_1	g_2
Value	1.1	0.7	0.1	0.2

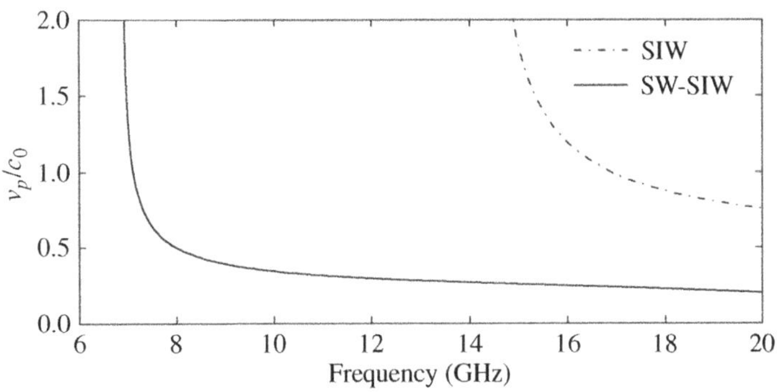

Figure 2.30 Simulated normalized phase velocity of the final SW-SIW and the normal SIW.

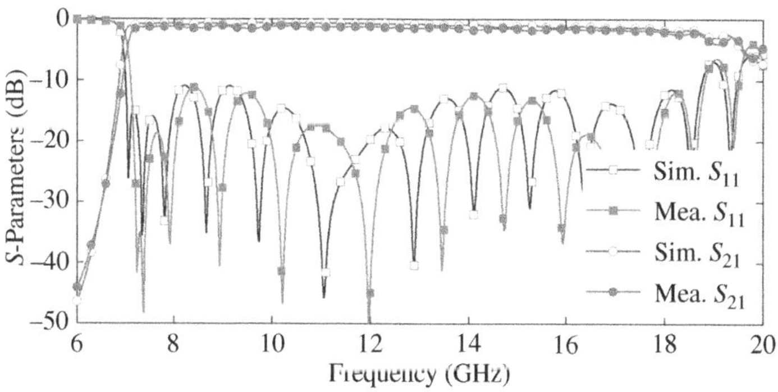

Figure 2.31 Simulated and measured S-parameters of the final 35-mm-long SW-SIW with the transition.

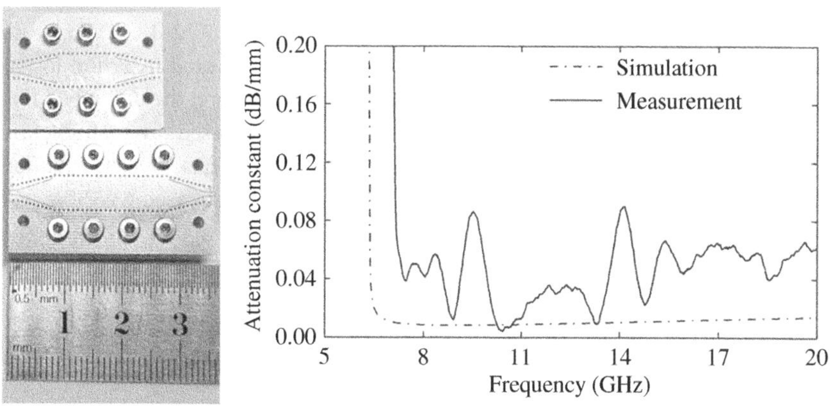

(a) (b)

Figure 2.32 (a) SW-SIW prototypes with different lengths of 26.6 and 35 mm. (b) Simulated and measured attenuation constant.

The final SW-SIW's simulated S-parameters, shown in Figure 2.31, demonstrate that the reflection coefficient stays above −10 dB in the X- and Ku-bands, which span the frequency range of 7.0–18.8 GHz. 91.5% is the fractional bandwidth. With a longitudinal length of 35 mm, the SW-SIW has an insertion loss in the X- and Ku-bands of 0.6 dB at the least and 1.8 dB at the maximum. Figure 2.32 shows the proposed SW-SIW's simulated attenuation constant, which varies from 0.008 to 0.015 dB/mm.

The Ansys HFSS eigenmode solver yields a Q factor of 207.1 at 7.08 GHz, which may be used to assess the loss of the SW-SIW. Because of the concentration of electric and magnetic fields around the SWSs, which results in larger dielectric and conductor losses, the insertion loss of the SW-SIW is comparatively higher than that of the standard SIW at the same physical longitudinal length. The electric and magnetic fields are more concentrated at higher SWFs, which leads to greater conductor and dielectric losses and, ultimately, a lower Q factor.

2.3.3.3 Measured Results

The Agilent PNA E8363C Vector Network Analyzer is used to measure the S-parameters of the SW-SIW with transition in the case of measured results. When the EM simulations and the observed S-parameters of the SW-SIW are compared in Figure 2.30, there is excellent agreement between them. The observed cutoff frequency, 7.12 GHz, deviates from the simulated value by a reasonable 1.7%. The air gap between the two substrates is responsible for the rise in the observed cutoff frequency because it lessens the capacitance impact between the top ground plane and the shorted patch. The observed operating band, which represents a 90.2% fractional bandwidth, is 7.12–18.82 GHz with S_{11} better than −10 dB. In the X- and Ku-bands, the measured insertion loss is 2.16 dB at the highest and 1.11 dB at the lowest. Furthermore, the de-embedding approach in reference [16], shown in Figure 2.32, is used to manufacture two SW-SIWs of varying lengths in order to recover the observed attenuation constant. Table 2.4 presents a comparison between the proposed SW-SIW and other existing SW-SIWs. It is clear from this comparison that the proposed SW-SIW provides the largest fractional bandwidth and the greatest degree of size reduction.

2.3.4 Conclusion

In summary, by concurrently raising the effective permeability and permittivity, a new SW-SIW with improved SWF and a wider bandwidth is suggested. The suggested SW-SIW achieves reduced longitudinal and lateral dimensions because of a greater SWF and a lower cutoff frequency, respectively, in contrast to the current SW-SIW that simply enhances effective permeability or permittivity. Furthermore, steady wave impedance along the periodic SWS gives the suggested SW-SIW a broader bandwidth. In particular, the longitudinal length is decreased by 73%, resulting in a lower longitudinal dimension for a given electrical length, while the lateral dimension is reduced by 53% in comparison to a typical SIW at the same cutoff frequency. With a fractional bandwidth of 91.5%, the operating bandwidth ($S_{11} \leq -10$ dB) spans the X- and Ku-bands, which range between 7.0 and 18.8 GHz. Additionally, a broadband switch from SW-SIW to GCPW is intended to make measurement verification easier. Measurements and manufacturing of a prototype of the suggested SW-SIW reveal excellent agreement with the calculations.

Table 2.4 Comparison of SW-SIW.

References	SW structure	Increasing permeability or permittivity	Effective permittivity	Effective permeability	Lateral size reduction (%)	Longitudinal size reduction (%)	Fractional bandwidth (%)
[8]	Blind via-hole	Permittivity	9.9	—	40	40	71.4
[17]	Blind via-hole	Permittivity	9.5	—	38	38	83.5
[9]	Microstrip polyline	Permeability	—	2.8	40	40	73.9
[10, 18]	Lumped inductor	Permeability	—	2.4	35	35	66.7
[18]	Ring mushroom	Permittivity	8.2	—	40	40	52.8
[19]	Antipodal blind via-hole	Permittivity	17.5	—	55	50	26.1
This work	Blind via-hole and shorted patch	Permeability and permittivity	14.6	1.5	53	73	91.5

References

1 Jiao, T.-Q. (2018). *The Application and Research of EBG (Electromagnetic Tape Gap) in Decoupling of Array Antennas*, Harbin Engineering University.

2 Rajo-Iglesias, E., Quevedo-Teruel, Ó., and Inclan-Sanchez, L. (2008). Mutual coupling reduction in patch antenna arrays by using a planar EBG structure and a multilayer dielectric substrate. *IEEE Transactions on Antennas and Propagation*, 56 (6), 1648–1655.

3 Zhang, J., Zhang, X.P., Shen, D.Y. et al. (2016). Gap waveguide PMC packaging for a SIW-GCPW-based filter. *IEEE Microwave and Wireless Components Letters*, 26 (3), 159–161.

4 Tayebpour, J., Ahmadi, B., Fallahzadeh, M. et al. (2019). A waveguide switch based on contactless gap waveguide technology. *IEEE Microwave and Wireless Components Letters*, 29 (12), 771–774.

5 Xiao, H.-B., Fang, C., and Lu, G.-Z. (2018). Theoretical analysis of the ideal gap waveguide structure. *Journal of Communication University of China (Natural Science Edition)*, 25 (3), 34–41.

6 Chen, J., Chen, C.-H., Wang, P.-Y. et al. (2021). Magic T design based on ridge gap waveguide. *Journal of Microwave*, 37 (S1), 77–80.

7 Bayat-Makou, N. and Kishk, A.A. (2018). Realistic air-filled TEM printed parallel-plate waveguide based on ridge gap waveguide. *IEEE Transactions on Microwave Theory and Techniques*, 66 (5), 2128–2140.

8 Niembro-Martín, A., Nasserddine, V., Pistono, E. et al. (2014). Slow-wave substrate integrated waveguide. *IEEE Transactions on Microwave Theory and Techniques*, 62 (8), 1625–1633.

9 Jin, H., Wang, K., Guo, J., Ding, S., and Wu, K. (2016). Slow-wave effect of substrate integrated waveguide patterned with microstrip polyline. *IEEE Transactions on Microwave Theory and Techniques*, 64 (6), 1717–1726.

10 Jin, H., Zhou, Y., Huang, Y.M., and Wu, K. (2017). Slow-wave propagation properties of substrate integrated waveguide based on anisotropic artificial material. *IEEE Transactions on Antennas and Propagation*, 65 (9), 4676–4683.

11 Deslandes, D. and Wu, K. (2006). Accurate modeling, wave mechanisms, and design considerations of a substrate integrated waveguide. *IEEE Transactions on Microwave Theory and Techniques*, 54 (6), 2516–2526.

12 Wu, K. (1999). Slow wave structures, in *Encyclopedia of Electrical and Electronics Engineering (on Microwave Theory and Techniques)* (ed. J.G. Webster), Wiley, New York, pp. 366–381.

13 Xu, F. and Wu, K. (2005). Guided-wave and leakage characteristics of substrate integrated waveguide. *IEEE Transactions on Microwave Theory and Techniques*, 53 (1), 66–73.

14 Savidis, I. and Friedman, E.G. (2009). Closed-form expressions of 3-D via resistance, inductance, and capacitance. *IEEE Transactions on Electron Devices*, 56 (9), 1873–1881.

15 Kazemi, R., Fathy, A.E., Yang, S., and Sadeghzadeh, R.A. (2012). Development of an ultra-wide band GCPW to SIW transition, in *IEEE Radio Wireless Symp*, Santa Clara, CA, pp. 171–174.

16 Mangan, A.M., Voinigescu, S.P., Yang, M.-T., and Tazlauanu, M. (2006). Deembedding transmission line measurements for accurate modeling of IC designs. *IEEE Transactions on Electron Devices*, 53 (2), 235–241.

17 Khalil, M., Kamarei, M., Jomaah, J., and Ayad, H. (2015). Compact wide-band and low loss substrate integrated waveguide. in *Proceedings of the 3rd International Conference on Technological Advances in Electrical, Electronics and Computer Engineering (TAEECE)*, Beirut, Lebanon (April 2015), pp. 130–134.

18 Li, X., Dhwaj, K, and Itoh, T. (2018). Single-layer slow-wave substrate integrated waveguide with enhanced capacitance. in *IEEE International Microwave Symposium Digest*, Philadelphia, PA, USA (June 2018), pp. 316–318.

19 Zhou, Y., Jin, H., Huang, Y.M. et al. (2020). Slow-wave effect enhanced substrate integrated waveguide with multi-antipodal blind via-holes and distributed metal strips. *IEEE Microwave and Wireless Components Letters*, 30 (8), 753–756.

3

Device Design Based on SW-SIW

3.1 Slow-Wave Substrate-Integrated Waveguide Miniaturized Phase Shifter

3.1.1 Introduction

In millimeter-wave and microwave circuits, phase shifters are crucial parts, providing fixed or adjustable phase shifts within a specific frequency band, and are widely used in phased array antennas [1–3], multi-beam antennas [4–6], and shift communication systems [7, 8], making them an indispensable component of the Butler matrix discussed in this paper. The Schiffman phase shifter, introduced in the 1950s [9, 10], is a classic and commonly used type, consisting of a folded coupled transmission line that achieves the desired phase shift by adjusting the coupling coefficient. While suitable for microstrip and coplanar waveguide implementations, Schiffman shifters are challenging to realize using substrate-integrated waveguide (SIW) technology. SIW phase shifters can be implemented by varying the waveguide's length or width, leading to delay line phase shifters [11, 12] and unequal-width phase shifters [13–16]. The slow-wave (SW) SIW implementation suggested in Chapter 2 serves as the foundation for the Butler matrix suggested in this work. Phase shifting is accomplished by changing the SW factor and the SW-SIW segment, taking advantage of the direct relationship between the SW factor and the phase shift constant of an SW transmission line. By altering the size of the SW structure inside the SW-SIW transmission line, the electromagnetic (EM) wave's phase shift constant can be modified, resulting in a 45° and linear phase shifter design. Compared to conventional SIW delay lines and unequal-width phase shifters, these proposed SW-SIW-based ones exhibit simpler designs with more compact sizes.

Slow-Wave Substrate-Integrated Waveguide: Designing of Devices and Antennas, First Edition.
Jing-Ya Deng, Jia-Yuan Yin, Fengxia Li, Yulong Liu, Li-Xin Guo, and Xiao-Hua Ma.

3.1.2 Theory of Phase Shifter

The role played by an individual phase shifter within this Butler matrix includes ensuring that there exists a fixed-phase difference in its output ports at both crossover junctions as well as at those connected with its respective quadrature coupler sections having −3-dB responses. Thus, not only does it become important but also crucially necessary that whenever such designed phase-shifting features are being simulated or tested for performance accuracy, they must take into account any inherent errors arising from these crossover junctions behavior(s). Given that each such crossover junction will have two isolated ports—where input signal(s) can only exit through one port—it becomes possible to simplify such junction analysis with lossless generality wherein we treat it simply as another form/exemplification of the standard transmission line model, which shall serve us here as our reference for comparing overall performances.

For any given transmission line, its phase shift is determined by

$$\varphi = \beta l \tag{3.1}$$

Here, β denotes the phase shift constant of the transmission line, and l is the length. For an SIW transmission line

$$\beta = \sqrt{\left(\frac{2\pi\sqrt{\varepsilon_r}f}{300}\right)^2 - \left(\frac{\pi}{W_{eff}}\right)^2} \tag{3.2}$$

Here, W_{eff} represents the effective width of SIW in millimeters, and the unit of f is GHz. The following equation can be used to obtain W_{eff}:

$$W_{eff} = W_{SIW} - 1.08 \cdot \frac{d^2}{p} + 0.1 \cdot \frac{d^2}{W_{SIW}} \tag{3.3}$$

By substituting Eq. (3.1) into (3.2), the phase shift is

$$\varphi = \beta l = l \cdot \sqrt{\left(\frac{2\pi\sqrt{\varepsilon_r}f}{300}\right)^2 - \left(\frac{\pi}{W_{eff}}\right)^2} \tag{3.4}$$

It is evident that two fundamental kinds of phase shifters may be produced by varying the SIW's length or width: (1) delay line phase shifters, which are achieved by altering the length, and (2) unequal-width phase shifters, which are achieved by altering the width. Figure 3.1 displays the two phase shifters' structural diagram.

For the phase shifter of the delay line, its phase shift is

$$\Delta\varphi = \varphi_{21} - \varphi_{43} = (l_1 - l_2)\sqrt{\left(\frac{2\pi\sqrt{\varepsilon_r}f}{300}\right)^2 - \left(\frac{\pi}{W_{eff}}\right)^2} \tag{3.5}$$

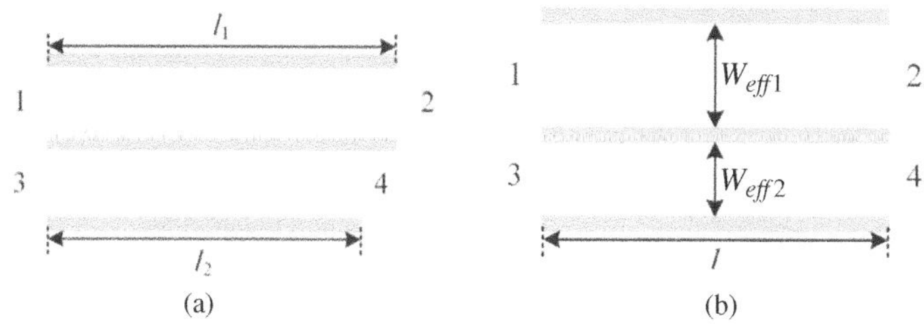

Figure 3.1 The phase shifter of (a) the delay line and (b) the unequal width.

And for the phase shifter of the unequal width, its phase shift is

$$\Delta\varphi = \varphi_{21} - \varphi_{43} = l\left(\sqrt{\left(\frac{2\pi\sqrt{\varepsilon_r}f}{300}\right)^2 - \left(\frac{\pi}{W_{eff1}}\right)^2} - \sqrt{\left(\frac{2\pi\sqrt{\varepsilon_r}f}{300}\right)^2 - \left(\frac{\pi}{W_{eff2}}\right)^2}\right) \tag{3.6}$$

With the goal of performing the phase shift function, it is evident that the phase shifter changes the waveguide width so as to change the phase shift constant β. The proposed SW-SIW phase shifter is based on the principle that the phase shifter of the unequal width varies the phase shift constant. Since the SW factor of an SW transmission line and the phase shift constant are positively connected, altering the SW factor of the SW-SIW may alter the phase shift constant.

3.1.3 Design of SW-SIW Phase Shifter

Ensuring a fixed phase difference between the 3-dB orthogonal coupler and the cross-junction's output port is the goal of the phase shifter in the Butler matrix. As a result, the cross-junction's phase shift must be taken into account and built into the suggested phase shifter. Figure 3.2 shows how the phase shifter is constructed. Two Rogers 5880 substrates are used. The thickness of the top dielectric substrate is h_1, and the thickness of the bottom dielectric substrate is h_2. The transverse width of the SW-SIW is denoted as W_0. The EM wave is confined within the SW-SIW by four rows of sidewall metalized via-holes, each with a diameter of d_2 and a transverse spacing of p between adjacent metalized through holes. Additionally, four rows of metalized blind holes with a diameter of d_1 are loaded on the lower dielectric substrate. On top of the bottom dielectric substrate, a rectangular short-circuit patch connected to the metalized blind holes is loaded.

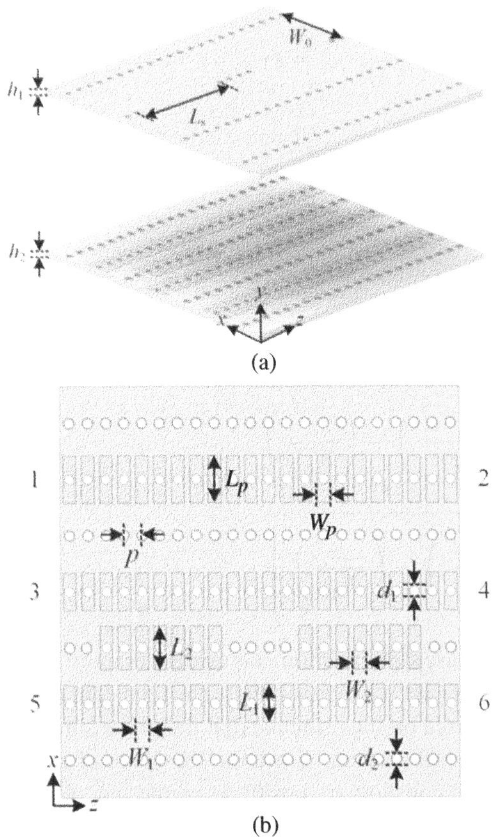

Figure 3.2 (a) Explosion diagram of SW-SIW phase shifter structure. (b) Diagram of the upper surface of the lower substrate shows two 3-dB cross-junctions cascaded in the middle and lower sides.

The dimensions of the cross-junctions are the same as those described in Section 3.2, with port numbers ranging from 3 to 6. An SW-SIW transmission line on the upper side serves as a phase shifter, and the phase shifter's short-circuit rectangular patch's width and length are indicated as L_p and W_p, respectively. By changing the SW structure's dimensions, the necessary phase shift may be obtained by altering the EM wave's equivalent relative permittivity and permeability and phase shift constant β. The phase shifter in this study is developed by adjusting the short-circuit rectangular patch's length, L_p. With the exception of the difference in L_p value, the design procedure for the 0° and 45° phase shifters in the Butler matrix is the same. However, the general principles remain the same. To illustrate the design process, we will focus on the 0° phase shifter.

When the length of the short-circuit rectangular patch, L_p, in the phase shifter is equal to the patch length L_1, in the cross-junction, the phase difference $\Delta\varphi$

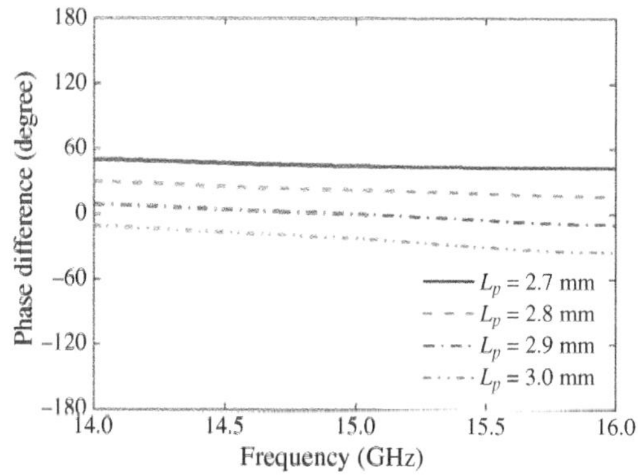

Figure 3.3 Port phase difference curves for different patch lengths.

between ports 2 and 6 at the 15 GHz design center frequency is 184°, as determined through full-wave software simulation. To achieve a phase difference of 0°, the phase of port 2 needs to be delayed by 184°. Via lengthening the phase shifter patch's length, L_p, distribution capacitance, equivalent relative dielectric constant, and EM wave phase shift constant β can be increased, thereby inducing a phase lag in port 2. Figure 3.3 illustrates the phase shift corresponding to different patch lengths, L_p, of the phase shifter. As observed, with an increase in the patch length, L_p, of the phase shifter, the EM wave's phase shift constant β increases, resulting in a greater phase lag in port 2 and a phase difference between port 2 and port 6 that approaches 0°. Finally, through simulation and optimization, the sizes of the SW-SIW 0° phase shifter are obtained, which are presented as follows.

The design procedure will not be repeated for the 45° phase shifter since it is the same as that for the 0° phase shifter. In Table 3.1, the SW-SIW 45° phase shifter's dimensions are determined via simulation and optimization.

Table 3.1 Dimensions of Slow-Wave Substrate-Integrated Waveguide 45° Phase Shifter Size.

Parameter	h_1	h_2	L_s	L_1	L_2
Value	0.508	0.508	7.86	2	2.35
Parameter	W_0	W_1	W_2	p	d_1
Value	6	0.7	0.7	1	0.5
Parameter	d_2	L_p	W_p		
Value	0.5	3.1	0.7		

3.1.4 Simulation Result

Figures 3.4 and 3.5 show the SW-SIW phase shifter's S-parameters and phase difference curves, respectively, that were created in this research. In the 14–16 GHz range, the phase difference between ports 2 and 4 is $0° \pm 15°$, the insertion loss is not less than 0.12 dB, and the reflection coefficient of the 0° phase shifter is better than −10 dB. The minimal insertion loss of the 45° phase shifter is 0.13 dB, a reflection coefficient is greater than −10 dB, and a phase difference is $45° \pm 15°$ between ports 2 and 4.

To demonstrate the downsizing impact of the proposed SW-SIW phase shifter, Figure 3.6 displays the design and dimensions of a conventional SIW phase shifter with an unequal width. W_9 is 15.3 mm for the 0° phase shifter and 17 mm for

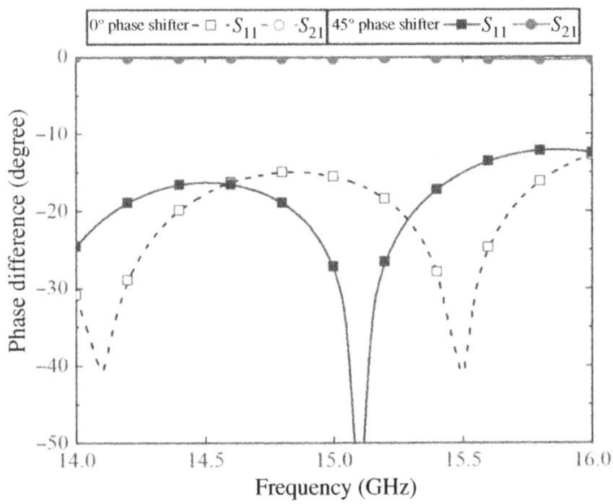

Figure 3.4 S-parameters of SW-SIW phase shifter.

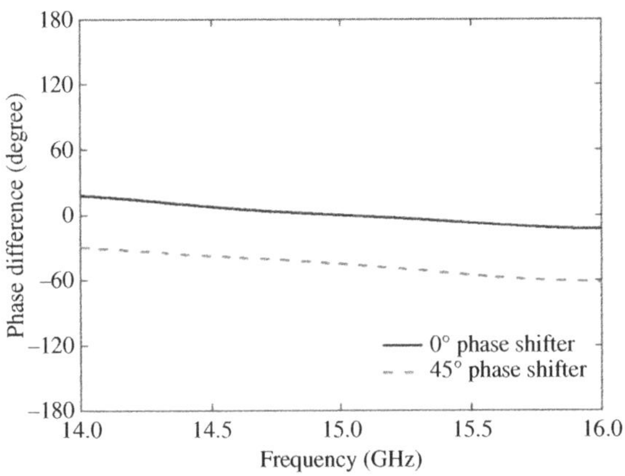

Figure 3.5 Phase difference curves of SW-SIW phase shifter.

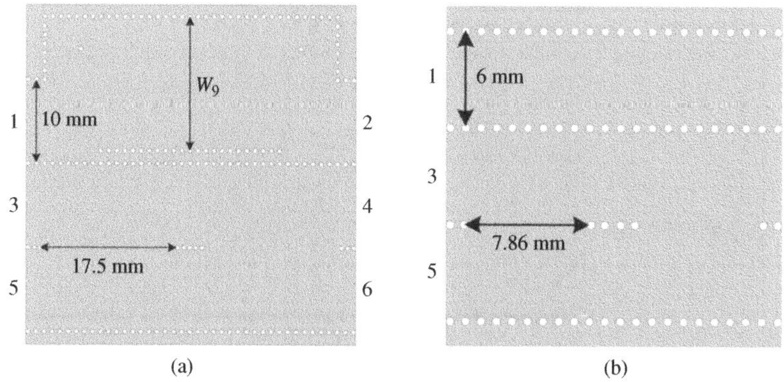

Figure 3.6 Diagram of (a) conventional SIW phase shifter and (b) SW-SIW phase shifter.

the 45° phase shifter. Figures 3.6 and 3.7 show the traditional SIW phase shifter's S-parameters and phase difference curves, respectively. In the 14.06–15.98 GHz frequency range, the insertion loss is at least 0.12 dB, the reflection coefficient is better than −10 dB, and the phase difference between ports 2 and 4 of the 0° phase shifter is 0° ± 15°. The 45° phase shifter features a minimal insertion loss of 0.11 dB, a reflection coefficient greater than −10 dB, and a phase difference of 45° ± 15° between ports 2 and 4. While the size of the SW-SIW 0° and 45° phase shifters is decreased by 74.4% and 75.6%, respectively, the performance is equivalent to that of the standard SIW phase shifters.

3.1.5 Conclusion

This chapter introduces the design of miniaturized devices based on SW-SIW transmission lines. Initially, a miniaturized 3-dB orthogonal short-slit coupler is developed using SW-SIW. By adding a short-circuit patch and a metalized blind hole, two SW structures, to the conventional SIW coupler, the coupler's transverse and longitudinal dimensions are decreased, while its equivalent relative permittivity and permeability are increased. The length of the coupling gap in the short-slit coupler is inversely proportional to the difference between the propagation constants of the TE_{10} and TE_{20} modes in the coupling area. The coupler's center and both sides have the strongest electric fields for the TE_{10} and TE_{20} modes, respectively. The longitudinal dimension of the coupler is further reduced by increasing the difference in propagation constants between the TE_{10} and TE_{20} modes by the amplification of the SW effect at the coupling gap. The SW-SIW couplers have 74% smaller dimensions than traditional SIW couplers while still performing similarly. Furthermore, the SW-SIW

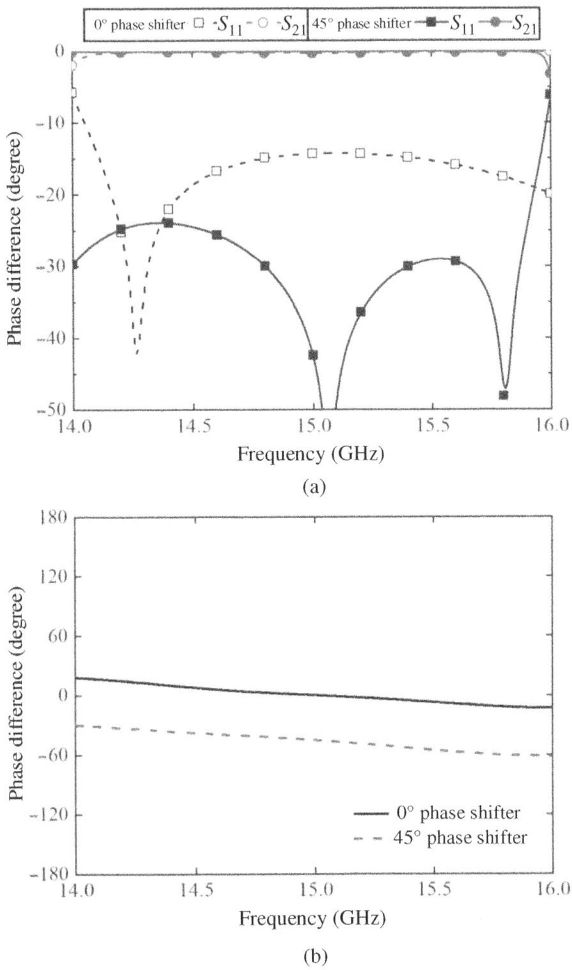

Figure 3.7 (a) *S*-parameters of conventional SIW phase shifter. (b) Phase difference curves of conventional SIW phase shifter.

couplers have a lower size in this study than those in earlier related work on SW-SIWs.

Subsequently, the 3-dB orthogonal short-slit coupler is used as the basis for designing an SW-SIW cross-junction. At the orthogonal coupler's pass-through port, signals with the same amplitude and orthogonal phase that are simultaneously input at the input and isolation ends cancel out in reverse, and at the coupling port, they join in phase. Consequently, two identical SW-SIW couplers are cascaded to create a miniature SW-SIW cross-connection. When compared to traditional SIW cross-junctions, the SW-SIW cross-junction is 70% smaller.

Lastly, by altering the phase shift constant of the EM wave and the size of the SW structure in the SW-SIW transmission line, 45° and 0° phase shifters are created. With a size decrease of 74.4% and 75.6%, respectively, the SW-SIW phase shifters operate similarly to traditional SIW phase shifters.

3.2 Slow-Wave Substrate-Integrated Waveguide Cross-Junction

3.2.1 Introduction

A cross-junction is a four-port device that facilitates the cross-transmission of EM energy while ensuring isolation between the two signal paths. It is a crucial component in a Butler matrix. Traditional microstrip cross-junctions can be implemented using structures such as air bridges [17], metallic rings [18], and branch lines [19]. However, the open structure of microstrip lines leads to significant radiation losses, which can degrade the system's electromagnetic compatibility (EMC) performance. SIW, as a novel transmission line offering low loss and easy integration, combines the advantages of metallic waveguides and microstrip lines. Cross-junctions based on SIW are easily integrated with planar circuits and exhibit minimal radiation loss. Moreover, as a critical element of the Butler matrix, the cross-junction must be miniaturized to reduce the size of the entire beamforming network. Therefore, this section designs a miniaturized cross-junction based on the SW-SIW transmission line proposed in Chapter 2.

3.2.2 Theory of Cross-Junctions

A cross-junction is a four-port device designed to achieve the cross-transmission of EM energy. Figure 3.8 shows the schematic representation of its concept. For a reciprocal and symmetric cross-junction, without loss of generality, Ports 2 and 4 are separated with no signal output; however, port 3 receives a signal that is input at port 1.

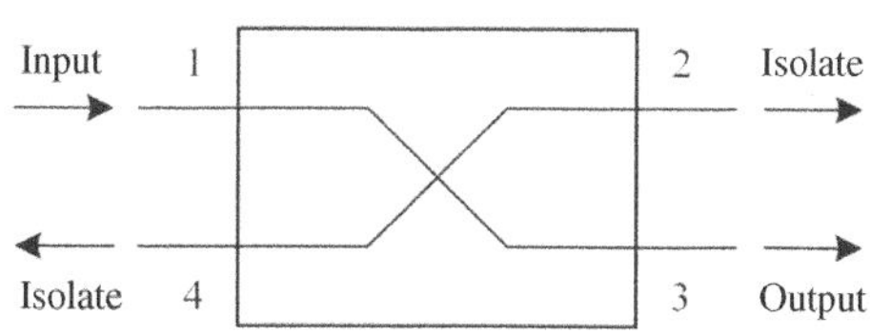

Figure 3.8 Cross-junction diagram.

A microwave device with four ports is called a cross-junction. Its matrix for the scattering parameter (S-parameter) is expressed as follows:

$$[S] = \begin{bmatrix} S_{11} & S_{12} & S_{13} & S_{14} \\ S_{21} & S_{22} & S_{23} & S_{24} \\ S_{31} & S_{32} & S_{33} & S_{34} \\ S_{41} & S_{42} & S_{43} & S_{44} \end{bmatrix} \tag{3.7}$$

For a reciprocal, symmetric, matched, and lossless passive cross-junction, the S-parameter matrix can be simplified as follows:

$$[S] = \begin{bmatrix} 0 & S_{12} & S_{13} & S_{14} \\ S_{12} & 0 & S_{14} & S_{13} \\ S_{13} & S_{14} & 0 & S_{12} \\ S_{14} & S_{13} & S_{12} & 0 \end{bmatrix} \tag{3.8}$$

According to the definition of a cross-junction, when a signal is input at port 1, it is output at port 3, while ports 2 and 4 are isolated. Therefore, $S_{12} = S_{14} = 0$, and $S_{13} = 1$. The resulting S-parameter matrix for the cross-junction is expressed as follows:

$$[S] = \begin{bmatrix} 0 & 0 & 1 & 0 \\ 0 & 0 & 0 & 1 \\ 1 & 0 & 0 & 0 \\ 0 & 1 & 0 & 0 \end{bmatrix} \tag{3.9}$$

$$[S] = \begin{bmatrix} 0 & 0 & 1 & 0 \\ 0 & 0 & 0 & 1 \\ 1 & 0 & 0 & 0 \\ 0 & 1 & 0 & 0 \end{bmatrix} \tag{3.10}$$

From Eq. (3.10), it is evident that a cross-junction possesses two isolated ports and demonstrates directionality. Essentially, it embodies a distinct form of directional coupler.

3.2.3 The Design Process of Slow-Wave Substrate-Integrated Waveguide (SIW) Cross-Junction

Considering that signals with the same amplitude and orthogonal phases are simultaneously input between the isolation and input ports of a 3-dB quadrature coupler, these two signals will cancel out in phase at the through port of the quadrature coupler and add up in phase at the coupled port. Additionally, these two input signals can be generated using a 3-dB quadrature coupler. Therefore,

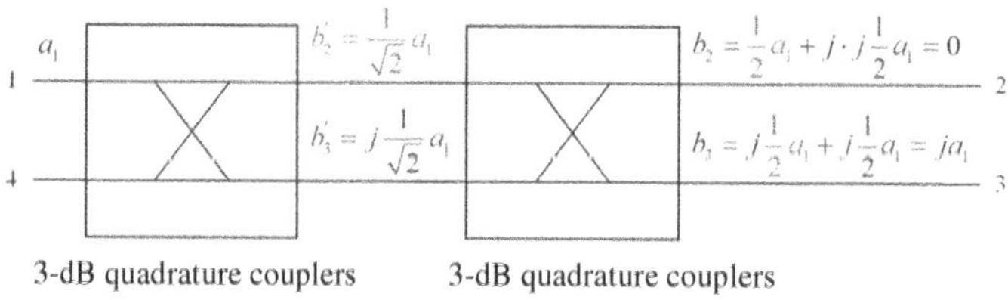

Figure 3.9 Cross-junction diagram of two couplers cascaded.

by cascading two identical 3-dB quadrature couplers, a cross-junction can be designed.

Figure 3.9 displays the schematic representation of the cross-junction that was created by cascading two identical 3-dB quadrature couplers. Without sacrificing generality, the output signals at the left 3-dB quadrature couplers through the port and coupled port are as follows:

$$b_3' = j\frac{1}{\sqrt{2}}a_1 \tag{3.11}$$

The output signals from the left coupler, b_2 and b_3, after they have passed through the right coupler, are added together at ports 2 and 3 of the cross-junction:

$$b_2 = \frac{1}{2}a_1 + j \cdot j\frac{1}{2}a_1 = 0 \tag{3.12}$$

$$b_3 = j\frac{1}{2}a_1 + j\frac{1}{2}a_1 = ja_1 \tag{3.13}$$

From Eq. (3.14), it is evident that when port 1 is energized, port 2 has no output, while port 3 outputs a signal with the same amplitude as port 1. Since port 4 of the cross-junction is the isolation port of the left 3-dB quadrature coupler, it also has no output signal. Therefore, the cross-junction can be realized by cascading two identical 3-dB quadrature couplers.

The structure of the SW-SIW cross-junction suggested in this study is shown in Figure 3.10. It consists of two SW-SIW couplers cascaded from Section 3.1, with dimensions matching those detailed in Table 3.2. The design incorporates two Rogers 5880 substrates with thicknesses h_1 and h_2 for the upper and lower substrate layers, respectively. W_0 is the SW-SIW's lateral width. EM waves are confined within the structure via metallization on three rows of sidewalls. The lateral distance between neighboring metalized through holes is p, and d_2 is the diameter of the sidewalls metallized through holes.

On the lower substrate layer, three rows of metalized blind holes with a diameter of d_1 are incorporated, along with short-circuit rectangular patches connected to these holes on the upper surface. The dimensions of the short-circuit rectangular patches flanking the coupling gap are designated as length L_1 and width W_1, while those within the coupling gap are labeled as length L_2 and width W_2.

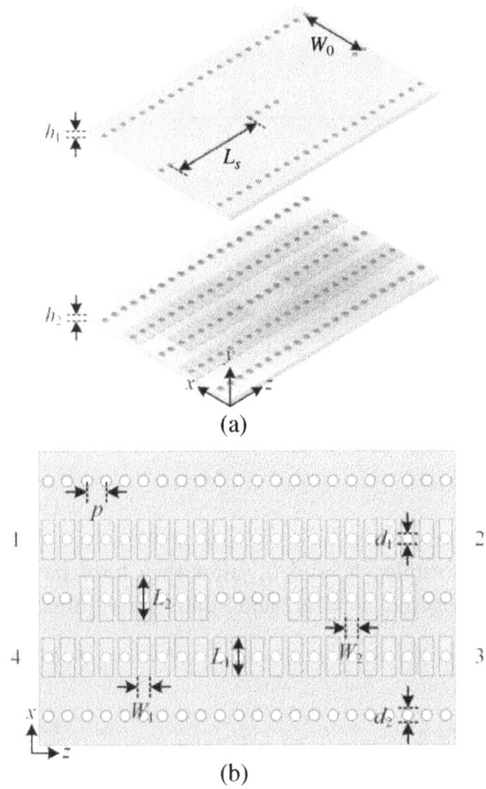

Figure 3.10 Structure diagram of slow-wave SIW cross-junction. (a) An exploded view in three dimensions. (b) Top view of the lower dielectric substrate.

Table 3.2 Dimensions of Slow-Wave Substrate-Integrated Waveguide 0° Phase Shifter Size.

Parameter	h_1	h_2	L_s	L_1	L_2
Value	0.508	0.508	7.86	2	2.35
Parameter	W_0	W_1	W_2	p	d_1
Value	6	0.7	0.7	1	0.5
Parameter	d_2	L_p	W_p		
Value	0.5	2.9	0.7		

Figure 3.11 displays the distribution of the H-plane electric field within the SW-SIW cross-junction when port 1 is excited. The input signal at port 1 is divided into two signals with the same amplitude and orthogonal phase by the left coupler. Following their passage via the appropriate coupler, these signals then merge in-phase at port 3 and cancel out at port 2. There is no signal output from port 4, which is the left coupler's isolation port.

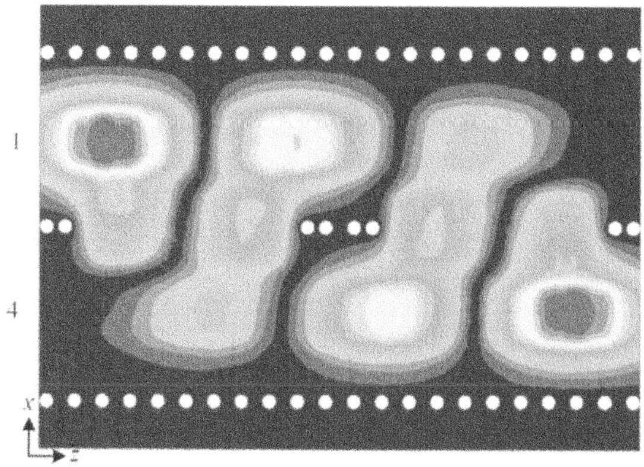

Figure 3.11 H-plane electric field distribution of slow-wave SIW cross-junction.

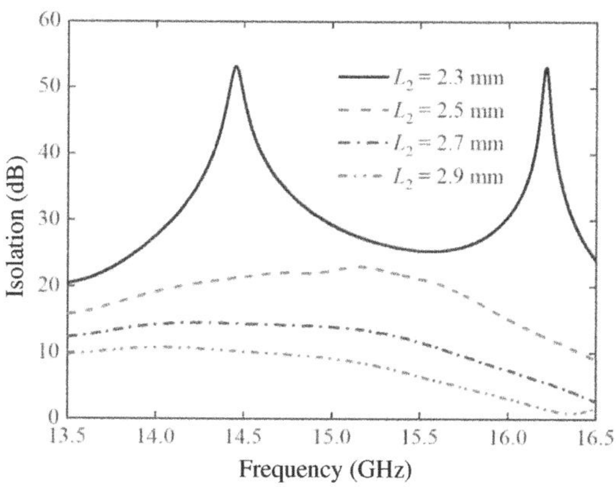

Figure 3.12 Effect of patch length L_2 on cross-junction isolation.

The SW-SIW cross-junction presented in this study consists of two identical 3-dB quadrature couplers connected in series. Therefore, the cross-junction's performance is directly impacted by the coupler's coupling coefficient. At the coupling gap, the length of the short-circuit rectangular patch influences the coupling coefficient of the coupler, as shown in Figures 3.15 and 3.16. When this coefficient is not 3 dB, it results in unequal amplitude and phase between output signals at port 2 of the right coupler, leading to a degradation in cross-junction isolation.

Figure 3.12 demonstrates that an increase in the length of the short-circuit rectangular patch at the coupling gap causes a deviation from a 3-dB coupling coefficient, subsequently decreasing isolation within the cross-junction. The H-plane field distribution corresponding to different patch lengths is depicted

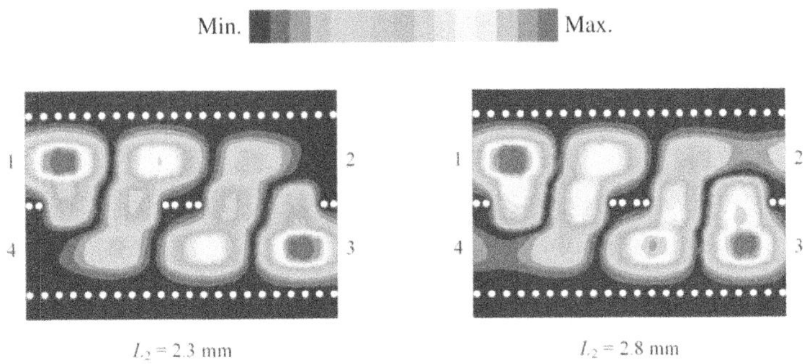

Figure 3.13 Distribution of the electric field in the cross-junction H-plane corresponding to different patch lengths.

in Figure 3.13, revealing that increasing patch length L_2 amplifies electric field amplitude at port 2 and consequently reduces isolation within the cross-junction.

3.2.4 Simulation Result

Figure 3.14 displays the S-parameter curves for the simulation results for this SIW cross-junction. In the 13.5–16.5 GHz frequency range, the negligible insertion loss is 0.18 dB, the isolation is above −10 dB, and the reflection coefficients are below −10 dB.

To emphasize the miniaturization effects achieved with our proposed SW-SIW cross-junction compared to conventional counterparts, we compared it with a

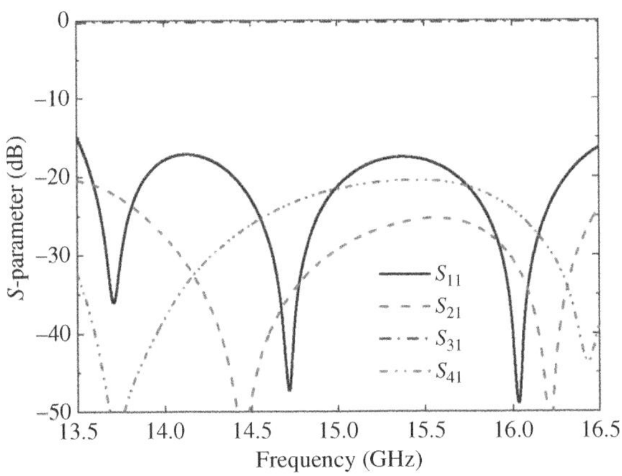

Figure 3.14 S-parameter curve of slow-wave substrate-integrated waveguide cross-junction.

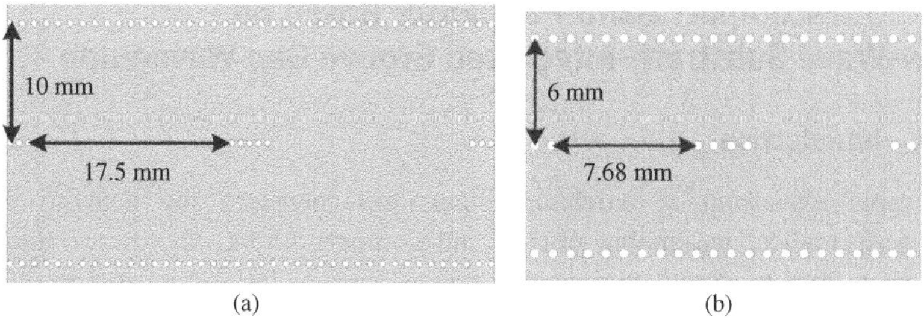

Figure 3.15 Comparison between ordinary SIW cross-junction and slow-wave SIW cross-junction (a) Traditional SIW cross-junction. (b) Slow-wave SIW cross-junction.

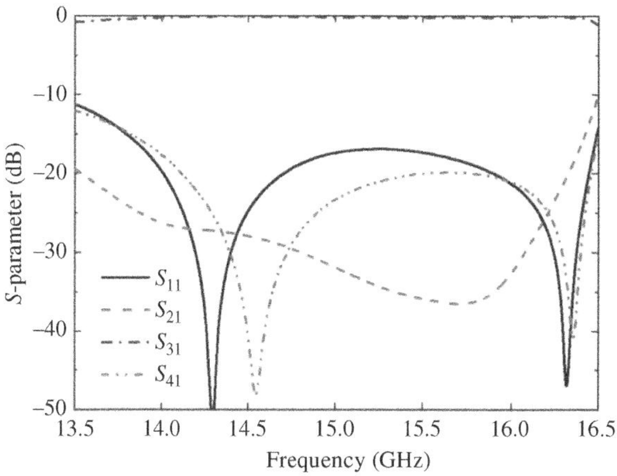

Figure 3.16 *S*-parameter curve of traditional SIW cross-junction.

conventional SIW cross-junction by cascading two identical conventional SIW couplers as outlined previously (Section 3.3).

Figure 3.16 shows the *S*-parameter curves for the traditional SIW cross-connection. In the 13.5–16.5 GHz frequency band, the reflection coefficient is better than −10 dB, the isolation is greater than 10 dB, and the minimal insertion loss is 0.14 dB.

In summary, the traditional SIW cross-junction and the SW-SIW cross-junction both operate similarly. The feasibility of miniaturizing the cross-junction utilizing the SW structure suggested in this study is validated by the 70% reduction in the size of the SW-SIW cross-junction when compared to the traditional SIW cross-junction.

3.3 Ultracompact Band-Pass Filter Based on Slow-Wave Substrate-Integrated Groove Gap Waveguide

3.3.1 Introduction

The rapid expansion of wireless systems has increased the demand for high-performance, reasonably priced, and compact filters. Microwave filters have been implemented using metallic hollow waveguides and microstrip lines [20, 21]. However, when the frequency rises, the microstrip experiences a significant loss, including metal, dielectric, and leakage loss. The large size, expensive price, and intricate production process of metallic hollow waveguides are its drawbacks.

Since its development in [22], the gap waveguide has been used more and more in millimeter-wave and microwave devices, including rotatable circuits [23, 24], antennas [25, 26], duplexers [27], couplers [28, 29], and filters [30, 31]. Because of its EM shielding structures, the gap waveguide has self-packaging properties that prevent energy leakage brought on by subpar electric connections. In order to package the guiding area, perfect magnetic conductors (PMCs) and paralleled perfect electric conductors (PECs) are used as EM shielding structures in the gap waveguide. The artificial magnetic conductor (AMC), which may be achieved by periodic components like metal pins [32] and mushrooms made on the substrate [33], often replaces the PMC. Metallic groove gap waveguide (GGW) [34], substrate-integrated groove gap waveguide (SIGGW) [35], ridge gap waveguide [36], and inverted microstrip gap waveguide [37] are among the several types of gap waveguides that have been created.

The benefits of high-Q factor, high power capacity, and self-packaging performance have led to the construction of metallic GGW filters [38, 39]. Nevertheless, these metallic GGW filters' large size restricts their use on small platforms. As a result, a number of GGW filters with smaller dimensions have been created. In [40], the resonators were stacked vertically to create a miniature GGW filter. In [41], a miniature GGW based on an SW GGW filter is created, where symmetrical capacitive quasiperiodic corrugations within the metallic GGW implement the SW characteristic. In comparison to the traditional metallic rectangular waveguide filter, the GGW filter's length is 57.1% shorter. However, since so many tiny periodic nails must be made, these full-metal GGW filters come at a high cost. The ultracompact band-pass filter (BPF) in this article is designed using SIGGW, a development of metallic GGW, which has the benefits of cheap cost and simple production.

This section proposes an ultracompact BPF with much decreased dimensions and self-packaging properties that are accomplished via SW-SIGGW. First, an SW-SIGGW transmission line is constructed in order to create a tiny GGW filter.

In contrast to the SW-SIGGW described in [42], the SW-SIGGW described in this article is made up of only two substrates: The top substrate realizes the gap layer, while the bottom substrate realizes the mushroom-type layer. In comparison to the standard SIGGW without an SW structure, the proposed SW-SIGGW can reach reductions of 51.6% and 52.8% in transverse and longitudinal dimensions, respectively, because of the SW structure in the shape of periodic metallic via-holes on the bottom substrate. This technique for reducing the waveguide's size is also confirmed by [43]. In addition to the metallic vias, additional patches on the blind via-holes are incorporated into the proposed structure. Such patches enable the simultaneous rise of permittivity and permeability, which produces a transmission line impedance that is comparatively steady. On the basis of the SW-SIGGW, a dual-band resonator and a cascaded second-order BPF are built. Comparing the cascaded second-order SW-SIGGW BPF to the cascaded BPF based on standard SIGGW, the decrease in area is 62.6%. The SW-SIGGW BPF is layered using a slot coupling structure and vertical stacking arrangement to achieve even more reduced dimensions, yielding fractional bandwidths of 5.4% and 7.4% at 11.8 and 18 GHz, an extremely competitive solution for applications with restricted space. In comparison to standard SIGGW BPF without SW and stacked structures, the final stacked SW-SIGGW BPF reduces the area by 81.3%.

3.3.2 Slow-Wave Substrate-Integrated Groove Gap Waveguide and Band-Pass Filter

The design procedure for the proposed SW-SIGGW BPF is illustrated in detail in this section. For the benefit of filter designers, it offers guidelines for the design and management of the filter's performance. Each design process has been divided into the following sections. (1) To build a filter, a SIGGW is shown in Section A as a self-packaged transmission line. (2) The SW-SIGGW was suggested in Section B. The filter may be made smaller by reducing the guided wavelength and the waveguide's lateral dimensions by the use of SW via-holes. (3) A miniature SW-SIGGW resonator was shown in Section C. (4) In Section D, an inductive window is used to adjust the coupling between resonators in a cascaded SW-SIGGW second-order filter. (5) An additional miniature stacked SW-SIGGW filter is suggested in Section E. The roll-offs caused by the transmission zeros, insertion losses, and coupling coefficients are discussed.

3.3.2.1 Substrate-Integrated Groove Gap Waveguide

The mushroom-type electromagnetic band gap (EBG) unit cell is initially created to accomplish the packaging structures of SIGGW, as shown in Figure 3.17. The

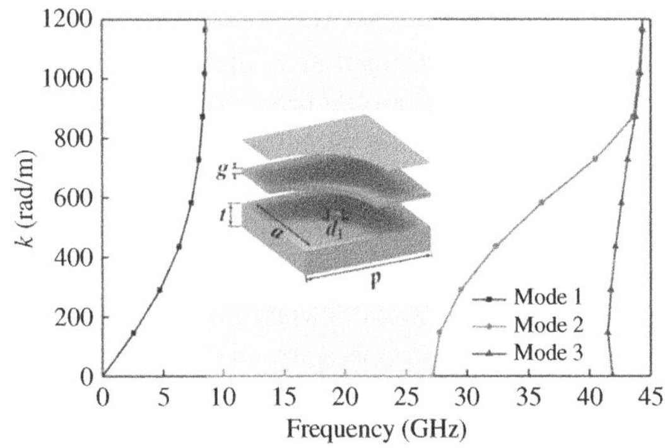

Figure 3.17 Dispersion diagram of the mushroom-type EBG unit cell.

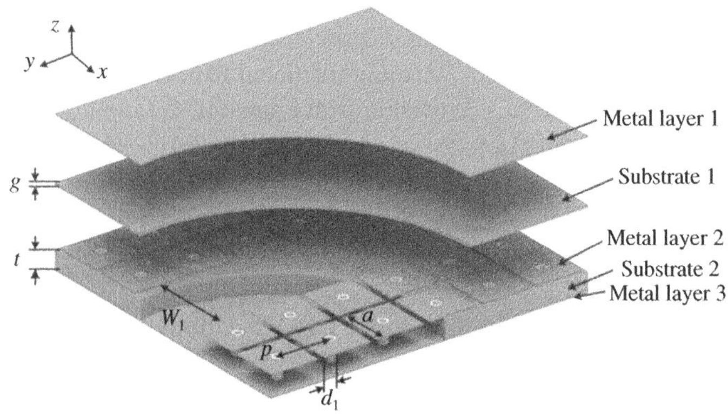

Figure 3.18 Structure of the SIGGW.

permittivity ε_r of both Rogers RT 5880 dielectric substrates is 2.2, and their loss tangent is $\tan\delta = 0.0009$. The specific measurements of the EBG unit cell of the mushroom type are as follows: When g is 0.127 mm and t is 0.787 mm, the thicknesses of the two substrate layers are selected. The cell p of the mushroom-type EBG unit has a period of 2.4 mm, and the diameter of the via-hole is fixed to d = 0.3 mm. The metal patch's dimensions are fixed at 2.1 mm. Figure 3.17 also displays the dispersion diagram of the given mushroom-type EBG, which was determined using the CST Eigenmode Solver. It is evident that the mushroom-type EBG's packaging structure can be utilized in the 8.6–27.2 GHz stopband between modes 1 and 2.

The geometry of the SIGGW, which consists of two substrate layers, is shown in Figure 3.18. Two sets of mushroom-shaped EBG unit cells are arranged on each side of the groove structure to serve as packaging structures. The port width W_1 of the SIGGW determines its cutoff frequency. The value of W_1 is fixed at 3.8 mm, and the cutoff frequency is 20 GHz. The attenuation per waveguide wavelength and the simulated S-parameters of the SIGGW are shown in Figure 3.19. The SIGGW attenuation is nearly constant across the whole operating band. It is evident that the transmission coefficient S_{21} is better than −0.3 dB from 21 to 30 GHz, while the reflection coefficient S_{11} is better than −15 dB spanning 20.98–39.03 GHz. The SIGGW attenuation is nearly constant across the whole operating band.

3.3.2.2 Groove Gap Waveguide Integrated with an SW Substrate

The SW-SIGGW shown in Figure 3.20 is the result of three rows of periodic metallic via-holes embedded in substrate 2 to increase relative permittivity and produce the SW effect for the standard SIGGW. $S_1 = 1.3$ mm and $S_2 = 1.25$ mm are the lateral and longitudinal distances, respectively, between two neighboring via-holes of the SW structure. The circular pads' diameter is set at $d_3 = 0.94$ mm, whereas the via-holes' diameter is $d_2 = 0.7$ mm. When choosing via-hole dimensions, there are primarily two factors to take into account. One is that a high degree of machining accuracy is not required for the size and placement of the via-holes in the cavity. The SW effect is the second, and it depends on the radius and period of the via-holes. In other words, the filter design becomes simpler as the period decreases

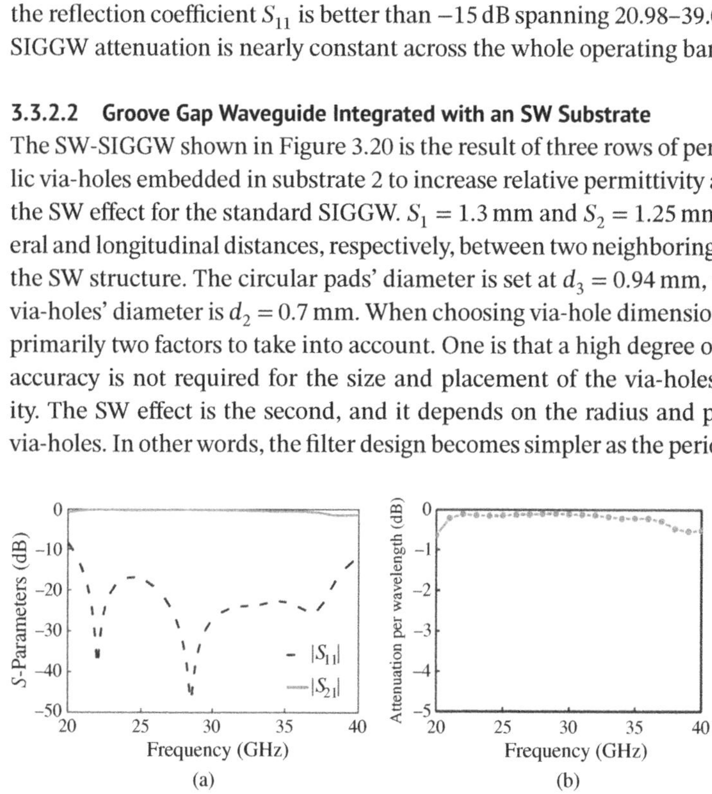

Figure 3.19 (a) S-parameters of the SIGGW. (b) Attenuation per wavelength of SIGGW.

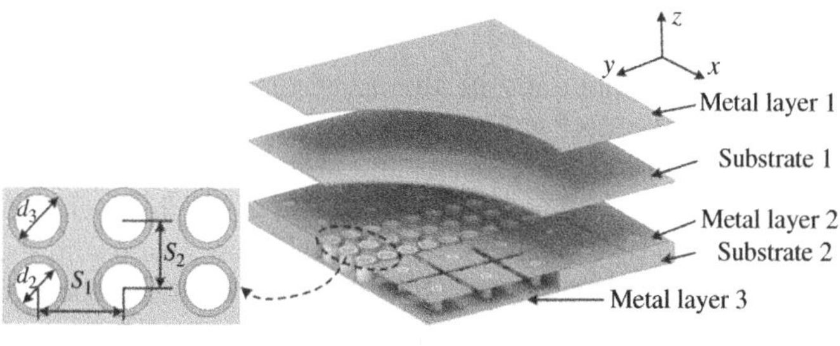

Figure 3.20 Configuration of SW-SIGGW ($d_2 = 0.7$ mm, $d_3 = 0.94$ mm, $S_1 = 1.3$ mm, and $S_2 = 1.25$ mm).

because the transmission line's distributed impedances vary less. The SW impact will be lessened, but by a brief interval. As a consequence, choosing the dimensions is essentially a trade-off between design and manufacturing viability and miniaturization (caused by the SW effect). It is evident that S_1 and S_2 vary somewhat from one another. Other than taking the filter's performance into account, this simply relies on the cavity's length and breadth. Some of the via-holes will extend beyond the cavity if S_2 is set to 1.3 mm, the same as S_1.

Slow-wave structures are used to concentrate the SW-SIGGW's electric field in substrate 1 while maintaining the spread of the magnetic field. The following formula may be used to determine the connection between the height ratio of the metalized via-holes t to the gap layer g and the effective relative dielectric constant ε_{reff} [44]:

$$\varepsilon_{reff} = \varepsilon_r \cdot \frac{t+g}{g} = \varepsilon_r \cdot \left(1 + \frac{t}{g}\right) \tag{3.14}$$

The f_{csw} (cutoff frequency) can be expressed as

$$f_{csw} = \frac{c_0}{2 \cdot W_{reff}\sqrt{\varepsilon_{reff}}} = \sqrt{\frac{\varepsilon_r}{\varepsilon_{reff}}} f_c \tag{3.15}$$

Additionally, it is possible to get the phase velocity versus frequency as

$$v_\varphi = \frac{c_0}{\sqrt{\varepsilon_{reff} \cdot \left(1 - \left(\frac{f_{csw}}{f}\right)^2\right)}} \tag{3.16}$$

Finally, the slow-wave factor (SWF) is defined as

$$\begin{aligned}\text{SWF} = \frac{v_\varphi}{c_0} &= \frac{1}{\sqrt{\varepsilon_{reff} \cdot \left(1 - \left(\frac{f_{csw}}{f}\right)^2\right)}} \\ &= \frac{1}{\sqrt{\varepsilon_{reff} \cdot \left(1 + \frac{t}{g}\right) - \varepsilon_r \cdot \left(\frac{f_c}{f}\right)^2}}\end{aligned} \tag{3.17}$$

As stated in [44], if the height ratio of the metalized via-holes t to the gap layer g increases, the SW effect will be amplified, leading to an elevated capacitance-to-inductance ratio and a considerable drop in the characteristic impedance of the SW-SIGGW. Therefore, it will be difficult to match the characteristic impedance of the SW-SIGGW to the 50-Ω feeding lines. As a result, the SIGGW must be traded off. Matching the SW-SIGGW's characteristic impedance to the 50-Ω feeding lines will thus be challenging. Therefore, a trade-off of the SIGGW must be made. The thicknesses of both the gap layer ($g = 0.127$ mm) and the SW structure layer ($t = 0.787$ mm) in this design can be obtained from the

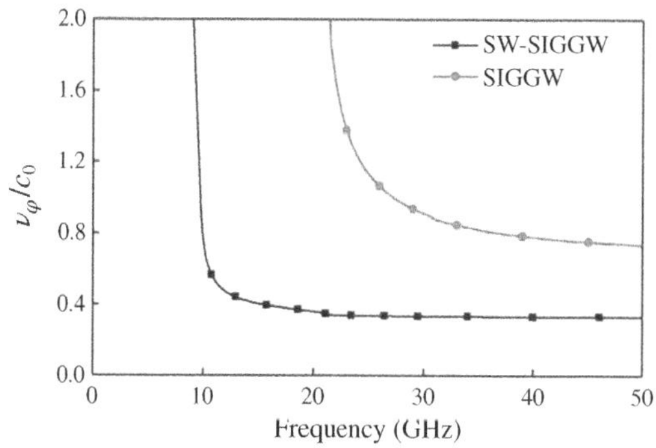

Figure 3.21 Phase velocity of SIGGW with/without slow-wave structure.

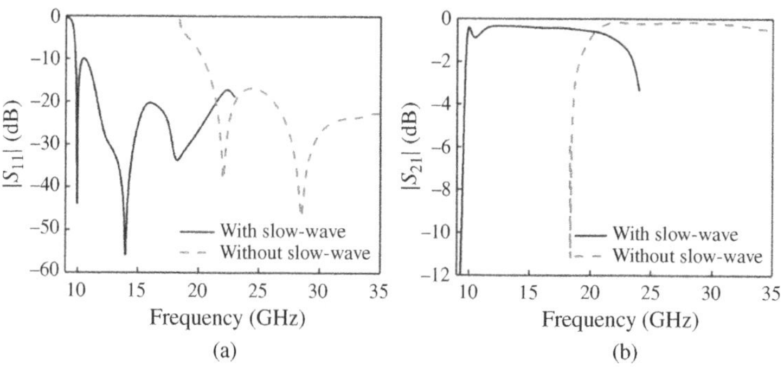

Figure 3.22 The simulated S-parameters of SIGGW. (a) $|S_{11}|$. (b) $|S_{21}|$.

supplier's catalog. The SW effect is demonstrated in Figure 3.21, which shows the SIGGW phase velocity as a function of frequency with and without SW structure, calculated by the CST Microwave Studio. It shows that, in comparison to standard SIGGW, the phase velocity of SW-SIGGW is 52.8% lower. Stated differently, SW-SIGGW may reduce longitudinal dimensions by 52.8% for a given electrical length when compared to regular SIGGW. Figure 3.22 displays the SIGGW's S-parameters with and without SW structure. It is evident that the SW-SIGGW's cutoff frequency is around 51.6% lower than that of regular SIGGW. Therefore, at the given cutoff frequency, incorporating the SW structure into traditional SIGGW may reduce the waveguide's lateral size.

3.3.2.3 The SW-SIGGW Resonator

The top view and 3-D schematic view of a novel cavity resonator constructed with the previously described SW-SIGGW are shown in Figure 3.23. Two shorting metalized via-holes are positioned on each side of the SW-SIGGW's port terminal to create a closed resonant cavity. The dimension values needed to achieve a closed

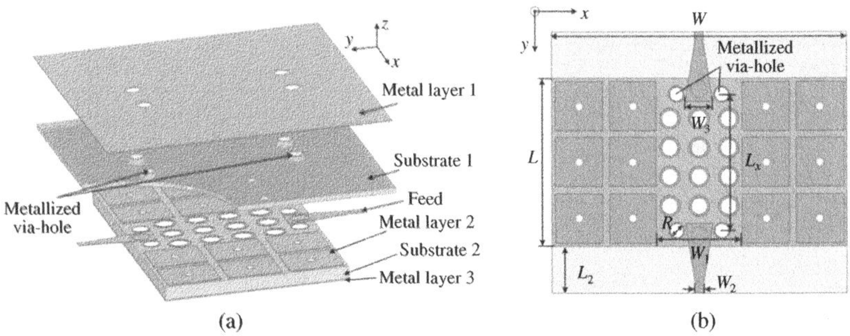

Figure 3.23 (a) Structure of the SW-SIGGW resonant cavity. (b) Top view of the SW-SIGGW resonant cavity ($W = 13.1$ mm, $L = 9.2$ mm, $W_2 = 0.4$ mm, $W_3 = 1.2$ mm, $L_2 = 2$ mm, $R = 0.6$ mm, $L_x = 5.84$ mm, and $W_1 = 3.8$ mm).

resonant cavity are labeled in Figure 3.23b. In the top view of the resonator, substrate 1 is made to be semitransparent, and the metal layer 1 is hidden for clarity. The diameter of the shorting metalized via-holes is R, and the cavity's length is L_x. The resonance frequency of the SW-SIGGW cavity resonator may be obtained using the following equation [45], as its resonance mechanism is comparable to that of the rectangular cavity resonator (length = b, width = d, and height = h):

$$f_{mnl} = \frac{c_0}{2\sqrt{\mu_r \varepsilon_{reff}}} \sqrt{\left(\frac{m}{b}\right)^2 + \left(\frac{l}{d}\right)^2 + \left(\frac{n}{h}\right)^2} \tag{3.18}$$

Here, $b > d > h$, c_0 is the velocity of light, μ_r is the relative permeability, and ε_{reff} is the SW-SIGGW cavity's effective dielectric constant. Considering that the substrate is extremely thin, the transmission mode of the SW-SIGGW is the TE mode, and n is 0. The resonator's mode numbers are the natural integers m, n, and l. For SW-SIGGW, (3.18) can thus be condensed to the following form:

$$f_{m0l} = \frac{c_0}{2\sqrt{\mu_r \varepsilon_{reff}}} \sqrt{\left(\frac{m}{L_x}\right)^2 + \left(\frac{l}{W_1}\right)^2} \tag{3.19}$$

By adjusting the cavity's width and length, the resonant frequency may be readily adjusted, as shown by formula (3.19). While the cavity's width W_1 is constant at 3.8 mm in this configuration, the cavity resonator's resonant frequencies fall as the cavity's length L_x rises. When L_x is set to 5.84 mm, the simulated S-parameters of the resonator are displayed in Figure 3.24. The resonance frequencies of two passbands are centered at 11.8 GHz for the TE101 mode and 18.5 GHz for the TE102 mode separately. Figure 3.24 illustrates the two passbands of the resonator, which are 11.73–11.94 and 18.02–18. GHz, with 3-dB fractional bandwidths of 1.8% and 5.2%, respectively. For both passbands, an insertion loss of better than 0.8 and

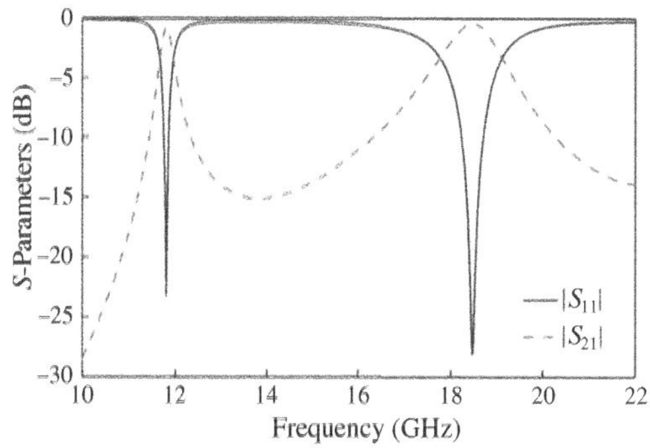

Figure 3.24 *S*-parameters.

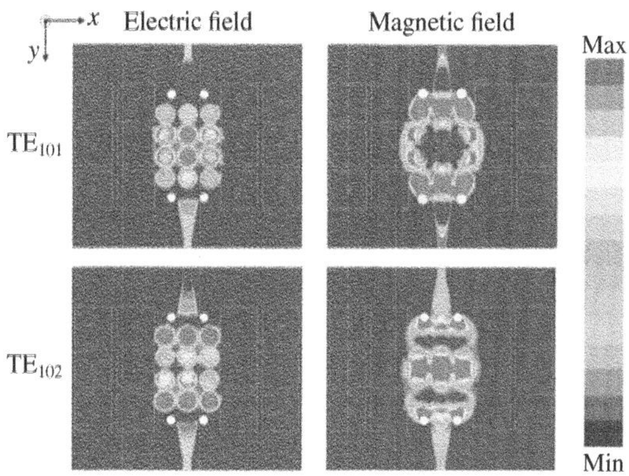

Figure 3.25 Distribution of the electric and magnetic fields for the TE101 and TE102 modes at 11.8 and 18.5 GHz, respectively.

0.4 dB is achieved. The first band of the TE_{101} mode determines the second band of the TE_{102} mode for the conventional waveguide. The addition of metallic vias to our design improves the frequency adjustment's versatility. The resonance frequency of TE_{102} can be customized by adjusting each of the metallic vias, which are located at the strongest field of TE_{102} and the weakest field of TE_{101}, separately. Consequently, the TE_{102} mode's field distribution is disturbed by these metallic vias, whereas the TE_{101} mode's field distribution remains largely unaltered. At the resonant frequencies of the resonator design based on SW-SIGGW, Figure 3.25 displays the electric and magnetic field distributions. At 11.8 and 18.5 GHz, the resonant modes TE_{101} and TE_{102} are clearly visible. As the SW cavity resonator's size varies, so does the suggested waveguide's quality factor. As shown in Figure 3.26, the quality factor is almost proportional to the resonator's length (L_x) as it grows.

3.3.2.4 The Cascaded SW-SIGGW Second-Order Filter

The second-order SW-SIGGW filter can be made by cascading two identical resonators, the shape of which is depicted in Figure 3.27a. W_l is the width of the

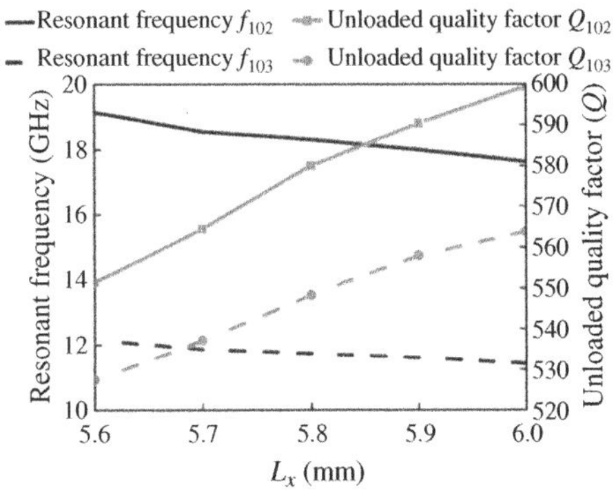

Figure 3.26 Quality factor against L_x.

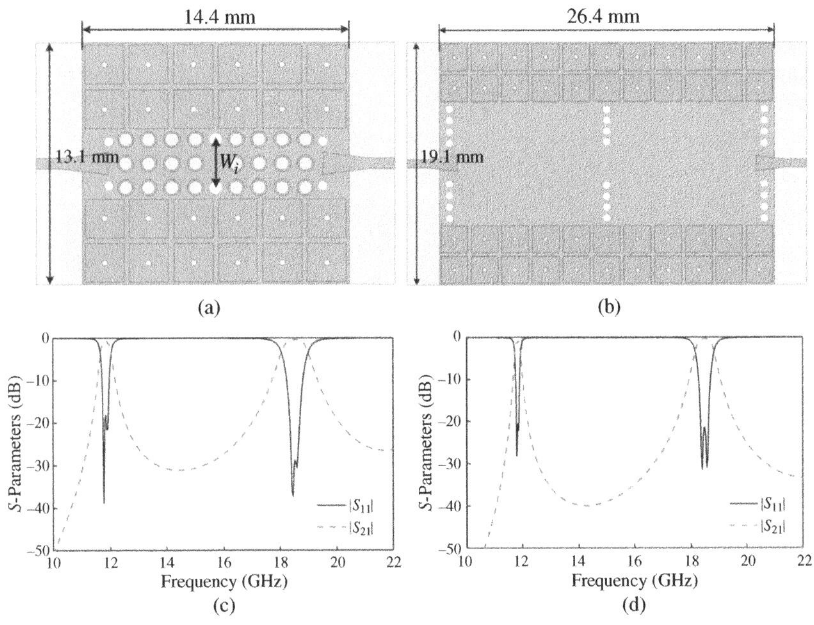

Figure 3.27 (a) The second-order SW-SIGGW filter's configuration. (b) The second-order SIGGW filter is configured without slow-wave structures. (c) The second-order SW-SIGGW filter's simulated *S*-parameters. (d) The second-order SIGGW filter's simulated *S*-parameters.

inductive window that realizes the coupling between two resonators. For comparison, Figure 3.27b also shows the setup of the second-order SIGGW filter without SW structures. In Figure 3.27c, W_i is 2.6 mm and the second-order SW-SIGGW filter's simulated *S*-parameters are shown. The SW-SIGGW filter has insertion losses of less than 0.8 and 0.5 dB and return losses of better than 18 dB when operating at 11.8 and 18.5 GHz. The two passbands have 3-dB fractional bandwidths of 5.3% (18.03–19.01 GHz) and 3.5% (11.63–12.05 GHz), respectively.

When SW vias are introduced, the waveguide's equivalent relative dielectric constant rises, which lowers the EM wave's velocity. The operating frequency shortens the EM wave's wavelength. Since the wavelength of EM waves determines the size of microwave devices, miniaturized devices are realized using the SW structure. This specific design uses metalized vias to build the SW structure. The wave impedance of the waveguide is reduced by the SW periodic vias' enhanced effective dielectric constant. Our design incorporates a transition from an SW waveguide to a feeding microstrip, which serves as an impedance transformer to prevent reflection.

Figure 3.27b shows a standard cascaded second-order SIGGW filter without an SW structure to highlight the amazing downsizing effect made possible by the SW structure. A typical second-order SIGGW filter measures 19.1 mm by 26.4 mm. In contrast, the SW-SIGGW filter has an area that is 62.6% smaller than that of a typical second-order SIGGW filter and measures 13.1 mm by 14.4 mm. Since the SW structure of the design, which is made of metallic vias, has a lower cutoff frequency, the size was decreased from 19.1 to 13.1 mm. Equations (3.14) and (3.15) show the cutoff frequency decreases when the SW periodic vias increase relative dielectric constant. For a given working frequency, reduced lateral dimensions may be achieved when the cutoff frequency drops.

Figure 3.27d displays the simulated S-parameters of the standard second-order SIGGW filter, and their filtering response is similar to that of the SW-SIGGW filter, demonstrating the SW structure effectively miniaturizes the filter in Figure 3.27c. However, the distance between the top plate and the top of the via-holes is reduced when metallic vias are introduced, which means that the electric field is focused more on the upper plate. For a given input power, power handling capacity decreases as the strength of the electric field increases. Additionally, S-parameters under various W_i's are calculated to further demonstrate how to regulate the coupling between the resonators. The connection between resonant frequencies and coupling coefficient k can be described as [46].

$$k = \frac{f_1^2 - f_2^2}{f_1^2 + f_2^2} \tag{3.20}$$

Here, f_1 and f_2 are the two resonant frequencies. According to formula (3.20), the coupling coefficient rises as the distance of two resonant frequencies goes further apart. Meanwhile, the bandwidth will increase with the distance of two resonant frequencies. Consequently, it may be said that the bandwidth increases with the coupling coefficient. As W_i increases, so does the bandwidth, according to the simulated findings shown in Figure 3.28. Therefore, the coupling between the two resonators becomes stronger as the width W_i grows.

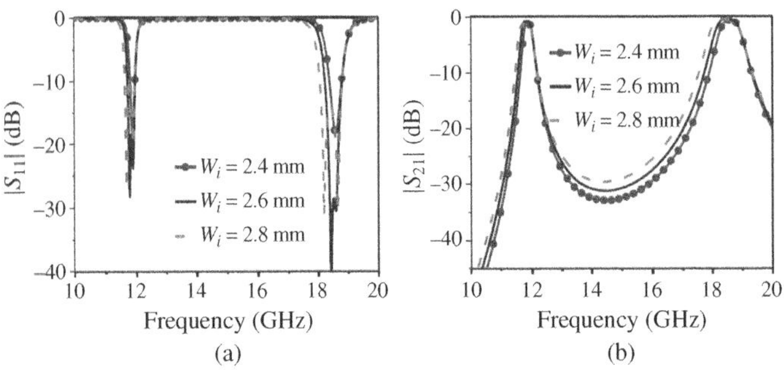

Figure 3.28 (a) $|S_{11}|$ and (b) $|S_{21}|$ against varied W_i.

3.3.2.5 Design and Optimization of the Stacked SW-SIGGW Filter

Figure 3.29 illustrates the geometry of a stacked SW-SIGGW filter, which is designed using a vertically stacked structure with coupling slots so as to significantly minimize the SW-SIGGW filter's size. There are four substrates in the stacked SW-SIGGW filter. Substrates 1 and 2 realize the top resonator, while substrates 3 and 4 realize the bottom resonator. Metal layer 3 is a metal that is shared by substrates 2 and 3. To provide the connection between the top and lower resonators, three slots are etched onto the metal layer 3. Two identical slots, designated slot B, measuring $px_2 \times py_2$, are carved symmetrically next to slot A, which has a $px_1 \times py_1$ dimension in the middle.

Figure 3.25 shows that the TE_{101} mode's electric and magnetic fields are localized around slot A and slot B, respectively. Thus, slot A predominantly affects the electric coupling, while slot B mostly impacts the magnetic coupling. Slot B determines the electric coupling for the TE102 mode, while slot A determines the magnetic coupling. The electric field concentrates around slot B, while the magnetic field concentrates around slot A. Transmission zeros may be produced for both TE_{101} and TE_{102} modes by the coexistence of electric and magnetic couplings, which can provide dual-coupling routes between two nearby resonators [47].

The connection between transmission zero f_z and resonant frequency f_o, as well as the coupling coefficient k between two resonators, can be described as [21].

$$k = \frac{f_{02}^2 - f_{01}^2}{f_{02}^2 + f_{01}^2} = \frac{M_c - E_c}{1 - M_c E_c} \tag{3.21}$$

$$\frac{f_z}{f_0} = \sqrt{\frac{E_c}{M_c}} \tag{3.22}$$

where M_c, E_c, and E_c are the magnetic and electric coupling coefficients, respectively, and f_{01} and f_{02} are the frequencies of the two transmission poles. As mentioned in [21], M_c and E_c of the mixed coupling may be derived. It is evident from

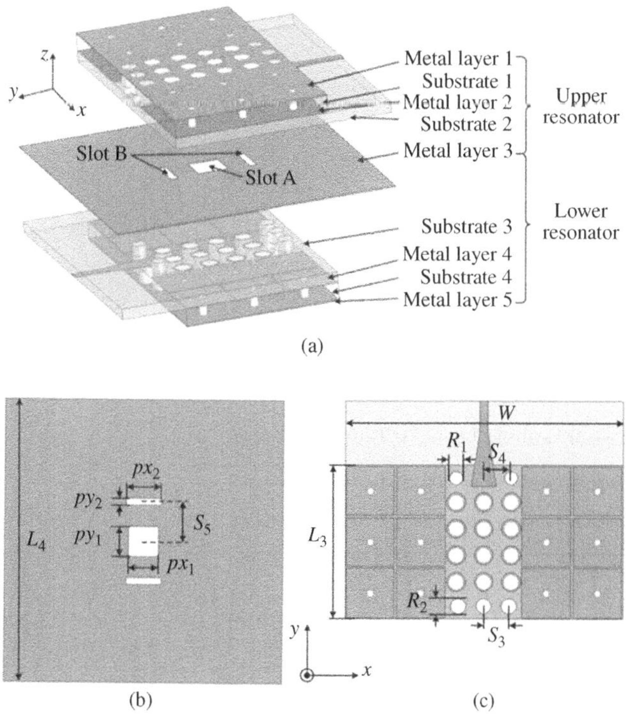

Figure 3.29 (a) Three-dimensional diagram of the stacked SW-SIGGW filter ($L_3 = 7.2$ mm, $L_4 = 9.2$ mm, $R_1 = 0.4$ mm, $R_2 = 0.7$ mm, $S_3 = 1.2$ mm, $S_4 = 1.25$ mm, $S_5 = 1.82$ mm, $py_1 = 1.2$ mm, and $py_2 = 0.3$ mm). (b) Layer 3. (c) The top view of the lower resonator.

(3.21) that the electric and magnetic couplings work together to determine the total coupling. As stated in (3.22), when the electric coupling is greater than the magnetic coupling, transmission zero is located above the passband. On the other hand, transmission zero appears below the passband when the electric coupling is weaker than the magnetic coupling.

Figures 3.30a and 3.31a show the changes of simulated $|S_{21}|$ in relation to slot A and slot B dimensions. M_c^{I} and M_c^{II} stand for the magnetic coupling coefficients for the first and second passbands, respectively. E_c^{I} and E_c^{II} stand for the electric coupling coefficients for the first and second passbands, respectively. f_z^{I} and f_z^{I} stand for the transmission zeros for the first and second passbands, respectively. Transmission zero f_z^{I} is clearly established above the first passband if $E_c^{\mathrm{I}} > M_c^{\mathrm{I}}$ and below the second passband if $E_c^{\mathrm{II}} < M_c^{\mathrm{II}}$. In both situations, when E_c/M_c rises, the transmission zero f_z^{I} moves to a higher frequency, and vice versa.

By altering the size of slots A and B, the stacked filter's filtering response in terms of selectivity and bandwidth may be changed. As shown in Figure 3.30, the

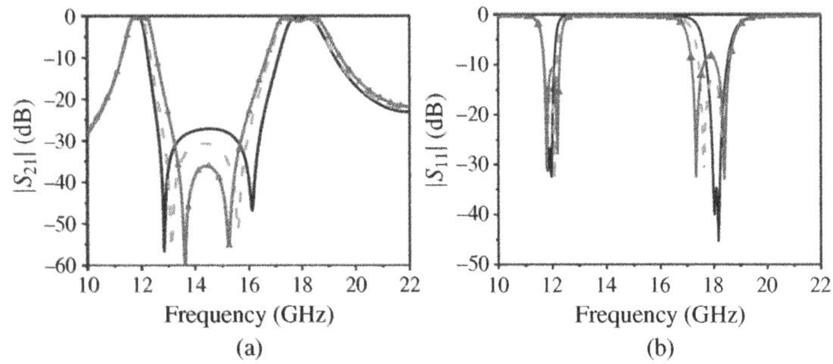

Figure 3.30 The stacked filter's (a) $|S_{21}|$ and (b) $|S_{11}|$ were simulated against varying px_1.

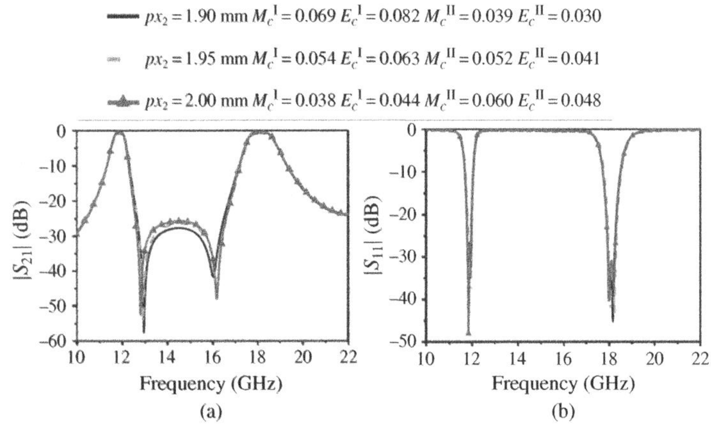

Figure 3.31 The stacked filter's (a) $|S_{21}|$ and (b) $|S_{11}|$ were simulated against varying px_2.

coupling becomes stronger as slot A px_1's size grows, resulting in two passbands with wider bandwidths. As a result, the two passbands' selectivity is reduced. Meanwhile, as Figure 3.31 illustrates, the bandwidth of the two passbands is somewhat reduced when the size of slot B px_2 grows because of the weaker coupling. As a result, two passbands have better selectivity. In order to obtain high performance of the stacked filter, the sizes of slots A and B are tuned as $px_1 \cdot py_1 = 1.7\,\text{mm} \times 1.2\,\text{mm}$ and $px_2 \cdot py_2 = 1.9\,\text{mm} \times 0.3\,\text{mm}$, respectively, taking into account the trade-off between bandwidth and selectivity. Table 3.3 lists the stacked filter's ultimate dimensions, and Figure 3.32 provides the appropriate simulated S-parameters of the stacked SW-SIGGW filter. As demonstrated, the 3-dB fractional bandwidths of the stacked filters for the two passbands operating at 11.8 and 18.0 GHz are 5.4% and 7.4%, respectively. There are two transmission

Table 3.3 Layout Dimensions of the Presented Stacked Filter.

Parameter	Value (mm)	Parameter	Value (mm)	Parameter	Value (mm)
W	13.1	L_3	7.2	L_4	9.2
R_1	0.4	R_2	0.7	S_3	1.2
S_4	1.25	S_5	1.82	px_1	1.7
py_1	1.2	px_2	1.9	py_2	0.3

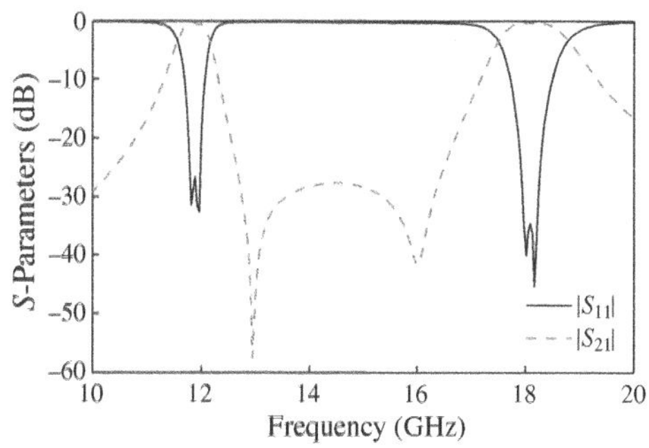

Figure 3.32 *S*-parameters of the stacked SW-SIGGW filter.

Table 3.4 Comparison of the Proposed Three Kinds of Filters.

Filters	Frequency (GHz)	FBW (%)	Size (mm × mm)	Size reduction (%)
Cascaded SIGGW filter	11.8/18.5	2.0/3.8	19.1 × 26.4	—
Cascaded SW-SIGGW filter	11.8/18.5	3.5/5.3	13.1/14.4	62.6
Stacked SW-SIGGW filter	11.8/18.0	5.4/7.4	13.1 × 7.2	81.3

zeros, f_z^{I} and f_z^{II}, at 12.9 and 16.0 GHz. The simulated return loss of the two passbands is better than 25 dB, while the insertion loss is less than 0.47 and 0.39 dB, respectively.

In Table 3.4, the filters discussed in this article are briefly compared. Comparing the developed second-order cascaded SW-SIGGW filter and the stacked SW-SIGGW filter to the cascaded SIGGW filter without SW structure, the two achieve 62.6% and 81.3% area reductions, respectively, confirming the miniaturizations that can be achieved by the SW structures and stacked configurations.

3.3.3 Fabrication and Measurements

The effectiveness of the developed SW-SIGGW BPF is verified using the built-in stacked second-order SW-SIGGW filter. The photographs of the fake BPF are shown in Figure 3.33. Tinol is used to solder substrates 2 and 3 together since they have etched coupling slots in common. Tinol's presence may result in an inevitable nonideal electrical contact, which would raise the insertion loss a little. Thick gaps between substrates 1 and 2 and substrates 3 and 4 are permitted without energy leakage, as the two resonators are constructed using SIGGW with self-packaging properties [48]. Substrates 1 and 2 and substrates 3 and 4 do not need bonding or soldering. Multiple layers are assembled and aligned using screws. Figure 3.34 compares the measured and simulated outcomes of the stacked second-order SW-SIGGW filter. According to the measurement, the two passbands' core frequencies are 12.0 and 18.2 GHz. The causes of the discrepancy between the measured and simulated findings have been examined. First, it is found that the thickness of the substrate affects the resonance frequency. In actual use, the substrate material may have thickness manufacturing errors. A Rogers 5880 substrate that is 0.127 mm thick has a maximum thickness inaccuracy of 0.0127 mm (±0.0005 in), according to the provider's data sheet. The simulated

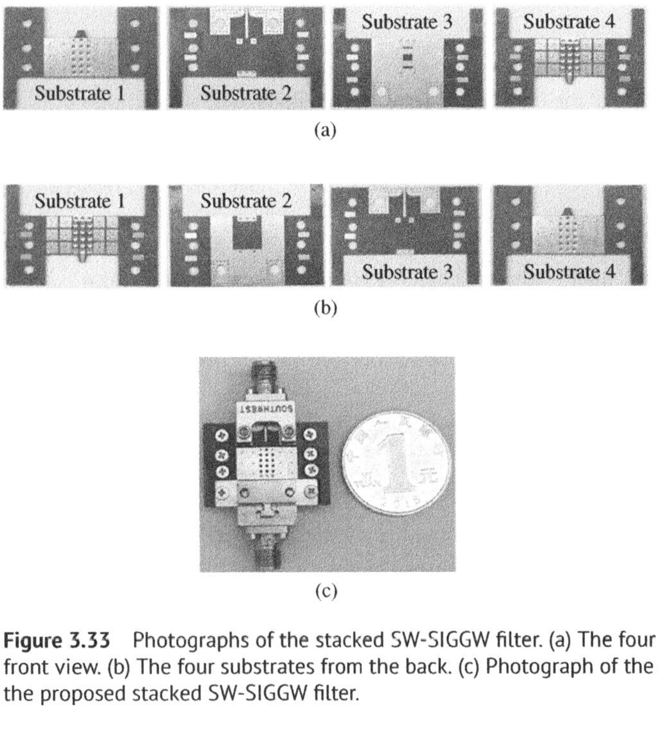

Figure 3.33 Photographs of the stacked SW-SIGGW filter. (a) The four substrates are in front view. (b) The four substrates from the back. (c) Photograph of the assembled form of the proposed stacked SW-SIGGW filter.

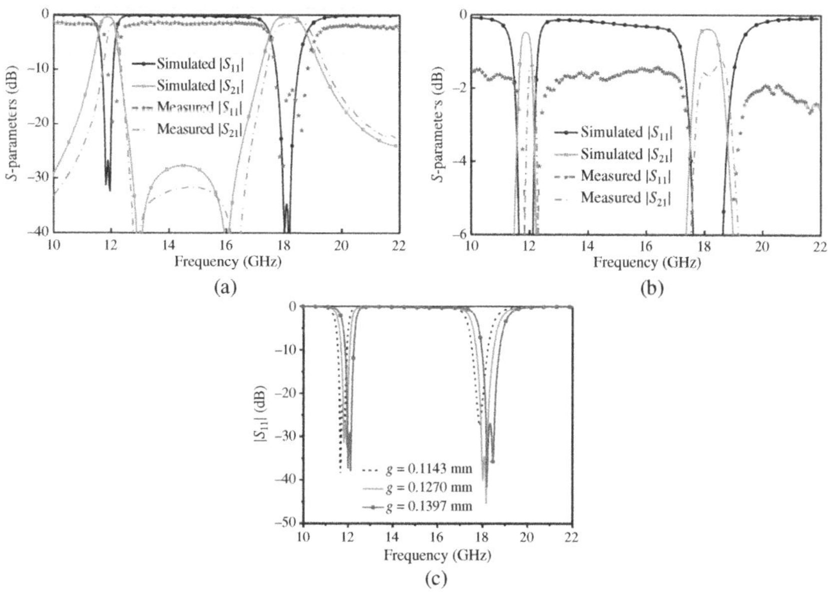

Figure 3.34 Simulated and measured *S*-parameters of the stacked SW-SIGGW filter. (b) A more detailed view of the dashed box in (a) and (c) simulated $|S_{11}|$ across a variety of substrate *g* heights.

passbands may move up or down due to the ±0.0127 mm thickness inaccuracy, as shown in Figure 3.34c. This explains why the *S*-parameters in Figure 3.34a differ between the simulated and measured values. Second, as shown in Figure 3.29, the splice between substrates 2 and 3 should be primarily responsible for the higher insertion loss. Perfect electrical contact holds these substrates firmly together in the simulation, leaving no gaps or additional structure. Tinol is used to solder the constructed substrates with the common ground in the measurement. An increase in insertion loss might be the consequence of the inevitable nonideal electrical contact.

Table 3.5 compares the performance of the suggested BPF with a number of published GGW BPFs. For comparison, an SW-SIW filter is also included [49]. The suggested SW-SIGGW filter exhibits a very compact volume and similar in-band performance to those GGW BPFs in [38–40, 49], according to the data shown in the table. Because of the dielectric loss present in the proposed SW-SIFFW BPF made by the dielectric substrates, the insertion loss (IL) of the SW-SIGGW filter is greater than that of the metallic GGW filter in [41]. The suggested SW-SIGGW filter is inexpensive and simple to integrate with planar circuits since it is made using PCB technology, while the other GGW BPFs in Table 3.5 are made using metal machining.

Table 3.5 Comparisons of Other GGW Filters with the Suggested Stacked Filter.

Reference	f_0 (GHz)	Ord.	FBW (%)	IL (dB)	Size ($\lambda_0 \times \lambda_0 \times \lambda_0$)	Material	ε_r	$\tan\delta$
[38]	14	5	1.0	1.22	1.11 × 4.72 × 0.34	Metal	/	/
[39]	35.65	4	1.0	1.2	2.43 × 2 × 2.43	Metal	/	/
[40]	35	3	1.0	1.0	2.08 × 1.88 × 1.48	Metal	/	/
[41]	11.87	6	3.88	0.5	2.64 × 2.31 × 0.38	Metal	/	/
[48]	11	5	8	3.2	1.00 × 0.24 × 0.04	Dielectric	3.55	5e−2
This work	**12/18.2**	**2**	**3.2/8.3**	**1.29/1.19**	**0.52 × 0.29 × 0.07**	**Dielectric**	**2.2**	**9e−3**

Notably, the frequency shown in the book is not for a specific application since the suggested design's working frequency may be adjusted by adjusting the resonators' diameters. The goal of this study is to provide a method for creating a small SW-SIGGW-based filter. The suggested method works regardless of the anticipated operation frequency. In other words, as the operating frequency varies, so do the applications for these filters.

3.3.4 Conclusion

This section proposes an ultracompact BPF based on a stacked configuration with SW-SIGGW. Two layers of the dielectric substrate make up the SW-SIGGW; the top substrate realizes the gap layer, while the bottom substrate realizes the mushroom-type layer. The SW structure, which takes the shape of periodic metallic via-holes on the bottom substrate, reduces the guided wavelength and cutoff frequency by about 52.8% and 51.6%, respectively, as compared to normal SIGGW. Such SW-SIGGW is used to create a cascaded second-order BPF and a dual-band resonator. The cascaded SW-SIGGW BPF reduces the area by 62.6% when compared to the normal second-order SIGGW BPF. By using a vertically stacked structure, the SW-SIGGW BPF is further reduced to 18.7%. The stacked SW-SIGGW BPF is constructed and measured; the findings demonstrate that the suggested SW-SIGGW BPF has a very small volume and is inexpensive, making it ideal for highly integrated and compact systems.

When compared to SIW or microstrip filters, the gap waveguide-based design has inherent benefits in self-packaging and noncontact. The suggested SIGGW filter has a benefit over microstrip filters in that it self-packages, preventing potential leakage loss. The suggested design has a noncontact advantage over SIW filters, which is advantageous for multilayer setups as it eliminates the requirement for excellent electric connections (such as flawless soldering) and may drastically lower manufacturing costs.

References

1 Toshev, A.G. (2008). Multipanel concept for wide-angle scanning of phased array antennas. *IEEE Transactions on Antennas and Propagation*, 56 (10), 3330–3333.

2 Miura, A., Fujino, Y., Taira, S. et al. (2005). S-band active phased array antenna with analog phase shifters using double-balanced mixers for mobile SATCOM vehicles. *IEEE Transactions on Antennas and Propagation*, 53 (8), 2533–2541.

3 Yin, L., Yang, P., Gan, Y.Y. et al. (2021). A low cost, low in-band RCS microstrip phased-array antenna with integrated 2-bit phase shifter. *IEEE Transactions on Antennas and Propagation*, 69 (8), 4517–4526.

4 Remez, J., Segal, A., and Shansi, R. (2005). Dual-polarized wideband widescan multibeam antenna system from tapered slotline elements array. *IEEE Antennas and Wireless Propagation Letters*, 4, 293–296.

5 Wu, Q., Hirokawa, J., Yin, J.X. et al. (2018). Millimeter-wave multibeam endfire dual-circularly polarized antenna array for 5G wireless applications. *IEEE Transactions on Antennas and Propagation*, 66 (9), 4930–4935.

6 Wincza, K. and Gruszczynski, S. (2016). Broadband integrated 8×8 butler matrix utilizing quadrature couplers and schiffman phase shifters for multibeam antennas with broadside beam. *IEEE Transactions on Microwave Theory and Techniques*, 64 (8), 2596–2604.

7 Liu, C.-H., Wei, R., and Zhou, H.-C. (2018). Performance analysis and simulation implementation of continuous phase modulation communication system. *Ship Electronic Engineering*, 38 (10), 93–97.

8 Wang, J. (2021). Research on the application of Continuous Phase Modulation (CPM) in tactical wireless communication. *China New Communications*, 23 (12), 1–2.

9 Schiffman, B.M. (1958). A new class of broad-band microwave 90-degree phase shifters. *IRE Transactions on Microwave Theory and Techniques*, 6 (2), 232–237.

10 Quirarte, J.L.R. and Starski, J.P. (1991). Synthesis of schiffman phase shifters. *IEEE Transactions on Microwave Theory and Techniques*, 39 (11), 1885–1889.

11 Xu, X.Y., Bosisio, R.G., and Ke, W. (2005). A new six-port junction based on substrate integrated waveguide technology. *IEEE Transactions on Microwave Theory and Techniques*, 53 (7), 2267–2273.

12 Tekkouk, K., Ettorre, M., and Sauleau, R. (2018). SIW rotman lens antenna with ridged delay lines and reduced footprint. *IEEE Transactions on Microwave Theory and Techniques*, 66 (6), 3136–3144.

13 Cheng, Y.J., Hong, W., and Wu, K. (2007). *Novel Substrate Integrated Waveguide Fixed Phase Shifter for 180-Degree Directional Coupler[C]//IEEE/MTT-S International Microwave Symposium*, IEEE, Honolulu, HI.

14 Morimoto, K., Hirokawa, J., and Ando, M. (2007). *Design of a 180-Degree Single-Layer Divider to Control Sidelobe and Crossover Levels in Butler-Matrix Beam-Switching Antenna[C]//Asia-Pacific Microwave Conference*, IEEE, Bangkok, Thailand.

15 Sbarra, E., Marcaccioli, L., Gatti, R.V. et al. (2007). *A Novel Rotman Lens in SIW Technology//European Radar Conference*, IEEE, Munich, Germany.

16 Cheng, Y.J., Hong, W., and Wu, K. (2009). Millimeter-wave multibeam antenna based on eight-port hybrid. *IEEE Microwave and Wireless Components Letters*, 19 (4), 212–214.

17 Zhou, J-J (1985). Research on "Air Bridge" [C]// Proceedings of the National Microwave Conference in 1985. 311–319.

18 Chiou, Y.C., Kuo, J.T., and Lee, H.R. (2009). Design of compact symmetric four-port crossover junction. *IEEE Microwave and Wireless Components Letters*, 19 (9), 545–547.

19 Kholodniak, D., G. Kalinin, E. Vernoslova. et al. (2000). Wideband 0-dB branch-line directional couplers. *IEEE MTT-S International Microwave Symposium Digest*, Boston, MA, 2000, 3: 1307–1310.

20 Iguchi, G., Tsuji, M., and Shigesawa, H. (2000). Negative coupling between TE10 and TE20 modes for use in evanescent-mode bandpass filters and their field-theoretic CAD, in *EEE MTT-S International Microwave Symposium Digest*, vol. 2, San Diego, CA, USA (May 1994), pp. 727–730.

21 Chu, Q.-X. and Wang, H. (2008). A compact open-loop filter with mixed electric and magnetic coupling. *IEEE Transactions on Microwave Theory and Techniques*, 56 (2), 431–439.

22 Kildal, P.-S., Alfonso, E., Valero-Nogueira, A., and Rajo-Iglesias, E. (2009). Local metamaterial-based waveguides in gaps between parallel metal plates. *IEEE Antennas and Wireless Propagation Letters*, 8, 84–87.

23 Sun, D., Chen, Z., Yao, C., and Xu, J. (2016). Flexible rectangular waveguide based on cylindrical contactless flange. *Electronics Letters*, 52 (25), 2042–2044.

24 Sun, D. and Xu, J. (2017). Real time rotatable waveguide twist using contactless stacked air-gapped waveguides. *IEEE Microwave and Wireless Components Letters*, 27 (3), 215–217.

25 Vukomanovic, M., Vazquez-Roy, J.-L., Quevedo-Teruel, O., Rajo-Iglesias, E., and Sipus, Z. (2016). Gap waveguide leaky-wave antenna. *IEEE Transactions on Antennas and Propagation*, 64 (5), 2055–2060.

26 Shi, Y., Zhang, J., Zeng, S., and Zhou, M. (2018). Novel W-band millimeterwave transition from microstrip line to groove gap waveguide for MMIC integration and antenna application. *IEEE Transactions on Antennas and Propagation*, 66 (6), 3172–3176.

27 Alós, E.A., Zaman, A.U., and Kildal, P.-S. (2013). Ka-band gap waveguide coupled-resonator filter for radio link diplexer application. *IEEE Transactions on Components, Packaging and Manufacturing Technology*, 3 (5), 870–879.

28 Ali, M.M.M., Shams, S.I., and Sebak, A.R. (2018). Printed ridge gap waveguide 3-dB coupler: analysis and design procedure. *IEEE Access*, 6, 8501–8509.

29 Shen, D., Wang, K., and Zhang, X. (2018). A substrate integrated gap waveguide based wideband 3-dB coupler for 5G applications. *IEEE Access*, 6, 66798–66806.

30 Lenoir, P., Bila, S., Seyfert, F., Baillargeat, D., and Verdeyme, S. (2006). Synthesis and design of asymmetrical dual-band bandpass filters based on equivalent network simplification. *IEEE Transactions on Microwave Theory and Techniques*, 54 (7), 3090–3097.

31 Mokhtaari, M., Bornemann, J., Rambabu, K., and Amari, S. (2006). Couplingmatrix design of dual and triple passband filters. *IEEE Transactions on Microwave Theory and Techniques*, 54 (11), 3940–3946.

32 Polemi, A., Maci, S., and Kildal, P.-S. (2011). Dispersion characteristics of a metamaterial-based parallel-plate ridge gap waveguide realized by bed of nails. *IEEE Transactions on Antennas and Propagation*, 59 (3), 904–913.

33 Liang, L., Liang, C. H., Zhao, X. W., and Su, Z. J.. (2008). A novel broadband EBG using multi-period mushroom-like structure, in *Proceedings of International Conference on Microwave and Millimeter Wave Technology*, Nanjing, China (April 2008), pp. 1609–1612.

34 Rajo-Iglesias, E. and Kildal, P.-S.2010). Groove gap waveguide: a rectangular waveguide between contactless metal plates enabled by parallel-plate cut-off, in *Proceedings of the Fourth European Conference on Antennas and Propagation*, Barcelona, Spain (April 2010), pp. 1–4.

35 Oyedokun, T., Geschke, R., and Stander, T. (2017). Experimental characterization of planar groove gap waveguide and cavity, in *Proceedings of 47th European Microwave Conference (EuMC)*, Nuremberg, Germany (October 2017), pp. 436–439.

36 Fan, F., Yang, J., Vassilev, V., and Zaman, A.U. (2018). Bandwidth investigation on half-height pin in ridge gap waveguide. *IEEE Transactions on Microwave Theory and Techniques*, 66 (1), 100–108.

37 Deng, J.-Y., Li, M.-J., Sun, D., Guo, L.-X., and Ma, X.-H. (2020). Compact dual-band inverted-microstrip ridge gap waveguide bandpass filter. *IEEE Transactions on Microwave Theory and Techniques*, 68 (7), 2625–2632.

38 Zaman, A.U., Kildal, P.-S., and Kishk, A.A. (2012). Narrow-band microwave filter using high-Q groove gap waveguide resonators with manufacturing flexibility and no sidewalls. *IEEE Transactions on Components, Packaging and Manufacturing Technology*, 2 (11), 1882–1889.

39 B. Al-Juboori, Y. Huang, D. Klugmann, M. Hussein, and J. Zhou. (2017). Millimeter wave cross-coupled bandpass filter based on groove gap waveguide technology, in *Proceedings of 0th UK-Europe-China Workshop on Millimeter Waves and THz Technologies (UCMMT)*, Liverpool, UK (September 2017), pp. 1–4.

40 Ahmadi, B. and Banai, A. (2015). Direct coupled resonator filters realized by gap waveguide technology. *IEEE Transactions on Microwave Theory and Techniques*, 63 (10), 3445–3452.

41 Liu, Z., Deng, J.-Y., and Sun, D. (2019). Slow-wave groove gap waveguide bandpass filter. *IEEE Access*, 7, 52581–52588.

42 Deng, J.-Y., Wang, Z.J., Sun, D. et al. (2020). Slow-wave substrate integrated groove gap waveguide. *IEEE Microwave and Wireless Components Letters*, 30 (5), 461–464.

43 Zhang, Y. et al. (2021). Slow wave substrate-integrated waveguide with miniaturized dimensions and broadened bandwidth. *IEEE Transactions on Microwave Theory and Techniques*, 69 (8), 3675–3683.

44 Niembro-Martín, A. et al. (2014). Slow-wave substrate integrated waveguide. *IEEE Transactions on Microwave Theory and Techniques*, 62 (8), 1625–1633.

45 Chen, B.-J., Shen, T.-M., and Wu, R.-B. (2009). Dual-band vertically stacked laminated waveguide filter design in LTCC technology. *IEEE Transactions on Microwave Theory and Techniques*, 57 (6), 1554–1562.

46 Hong, J.-S. (2011). *Microstrip Filters for RF/Microwave Applications*, 2nd edn, Wiley, Hoboken, NJ.

47 Gong, K., Hong, W., Zhang, Y., Chen, P., and You, C.J. (2012). Substrate integrated waveguide quasi-elliptic filters with controllable electric and magnetic mixed coupling. *IEEE Transactions on Microwave Theory and Techniques*, 60 (10), 3071–3078.

48 Bayat-Makou, N. and Kishk, A.A. (2018). Contactless air-filled substrate integrated waveguide. *IEEE Transactions on Microwave Theory and Techniques*, 66 (6), 2928–2935.

49 Bertrand, M., Liu, Z., Pistono, E. et al. (2015). A compact slow-wave substrate integrated waveguide cavity filter, in *IEEE MTT-S International Microwave Symposium Digest*, Phoenix, AZ, USA (May 2015), pp. 1–3.

4

Antenna Design Based on SW-SIW

4.1 Horn Antenna with Reduced Size and Enhanced Gain by the Loading of Slow-Wave Periodic Metal Blocks

4.1.1 Introduction

This section introduces an H-plane horn antenna that loads slow-wave structures (SWSs) into the shorter horn cavity to achieve both higher gain and reduced size. The periodic metal blocks that make up the SWSs slow down the electromagnetic wave's phase velocity as it approaches the radiating aperture. Consequently, the horn antenna's longitudinal length may be greatly reduced by compensating for the phase difference between the center and margins of the aperture, which is magnified by the lowered longitudinal dimension. Furthermore, the heights of the embedded blocks are strategically arranged to maximize the realized gain and aperture efficiency by achieving an even distribution of electric field magnitudes across the aperture. It is simple to build the suggested horn antenna using conventional mechanical production techniques. The prototype's longitudinal length is 33.3% less than that of the initial horn antenna with the same size of aperture. Additionally, the aperture efficiency is optimized at 80.7%, and compared to the original horn, the average gain is 1.7 dB higher.

In this section, we present an H-plane horn antenna that loads SWSs to achieve a smaller size and higher gain. The SWS is composed of periodic metal blocks that are scattered over the inner H-walls of the horn antenna's flared region. Each block's height and form are meticulously planned to provide a comparatively even distribution of phase and amplitude on the horn aperture. Stepped-height transition structures are also incorporated around the radiating aperture and throat of the horn to improve impedance matching. Metal fabrication is used to create the suggested SW horn. With the same aperture size, the prototype's longitudinal length is 33.3% shorter than the original horn antennas, but its average gain is 1.7 dB more.

Slow-Wave Substrate-Integrated Waveguide: Designing of Devices and Antennas, First Edition.
Jing-Ya Deng, Jia-Yuan Yin, Fengxia Li, Yulong Liu, Li-Xin Guo, and Xiao-Hua Ma.

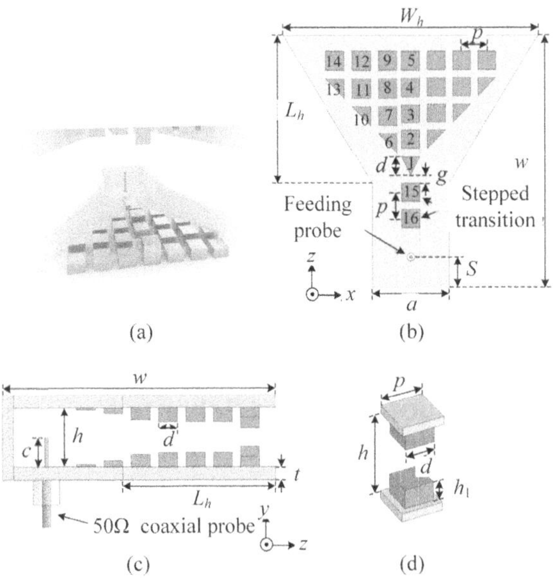

Figure 4.1 The suggested horn antenna's configuration. (a) 3-D view. (b) Top view. (c) Cross-sectional view. (d) Unit of the SWSs.

4.1.2 Antenna Design

Figure 4.1 displays the tiny horn antenna's geometry. At 12.5 GHz, the aperture measures $2.5\lambda_0 \times 0.57\lambda_0$. The integrated SWS blocks and the H-plane horn antenna make up the antenna's two primary components. The precise design parameters for the H-plane horn antenna are listed in Table 4.1. The original horn's and the shortened horn's designs, which both keep the same aperture size, are initially shown in the parts that follow. Next, we outline the SWS's theoretical analysis and design procedure. Lastly, we talk about how to make impedance matching better.

Table 4.1 Dimensions of the Designed Antenna.

Parameters	Value (mm)	Parameters	Value (mm)	Parameters	Value (mm)
h	9.53	W_h	50	d	3.5
a	15	s	5.8	t	2
L_h	28	p	5	c	4.765
w	50	g	1.5		

4.1.2.1 Shortened Horn

The shorter horn was modeled after the original horn antenna shown in Figure 4.2a. The flared section L_h is 42 mm long (equal to $1.75\lambda_0$), and the aperture width W_h is 50 mm (equivalent to $2\lambda_0$ at 12.5 GHz). This original design acts as a comparative counterpart to show that it is possible to achieve both improved gain and smaller size for a horn antenna at the same time. By reducing the flared area L_h to 28 mm (or $1.17\lambda_0$), we shrink the horn antenna and make it 33.3% shorter than the prior conventional horn. There is no change in the other dimension parameters. For comparison, a geometrical drawing of the regular horn and the reduced horn is shown in Figure 4.2. Reference [1] is a description of the phase difference ψ between the aperture's edge and the center.

$$\psi = \frac{\pi W_h(W - a)}{4\lambda L_h} \tag{4.1}$$

where W_h is the horn aperture width, a is the feeding waveguide width, and L_h is the horn's flared region length. If L_h is shortened and all other parameters remain the same, the gain will decrease because the aperture's phase difference will rise.

4.1.2.2 The SWSs Theoretical Analysis

An SWS is loaded in the flared area to provide a more homogeneous distribution of phase and magnitude on the aperture of the shorter horn. The corrugated waveguide served as an inspiration for the SWS's design [2–5]. Figure 4.1d displays the SWS unit's setup. The block-shaped SWS's height h_1 and side length d define the SW effect. An eigenmode simulation in CST is performed to demonstrate the operation of SWS. As shown in Figure 4.3a, when two face-to-face blocks are connected by an extended parallel capacitive effect, the propagation constant grows with block height. Figure 4.3b illustrates the connection between the block's width and the propagation constant. The relationship between the block's width and the propagation constant is shown in Figure 4.3b. When the effective permeability between two blocks side by side decreases, the propagation constant first

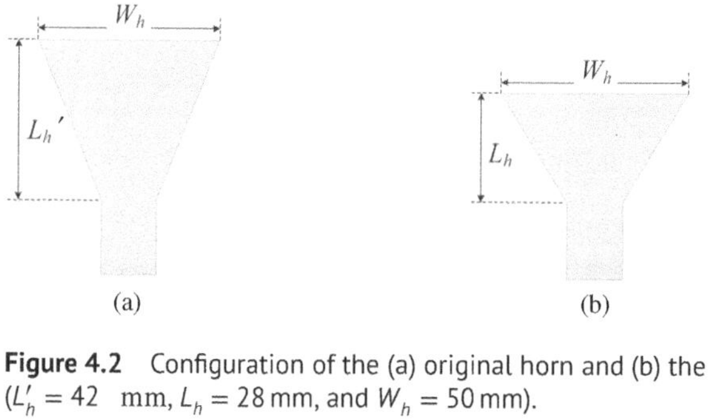

Figure 4.2 Configuration of the (a) original horn and (b) the shortened horn ($L_h' = 42$ mm, $L_h = 28$ mm, and $W_h = 50$ mm).

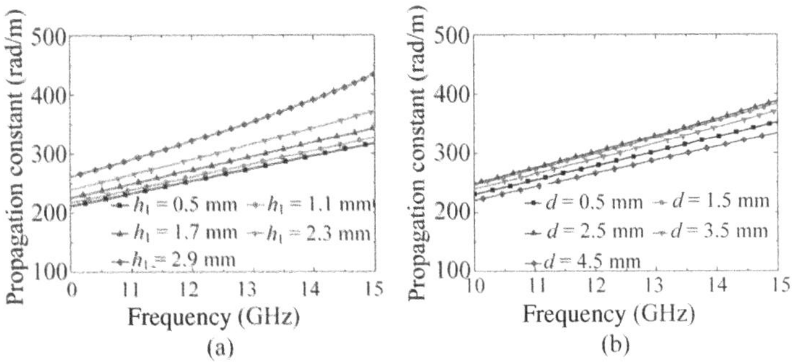

Figure 4.3 Simulations of dispersion diagrams for various SWS unit dimensions, (a) for varying heights h_1 and (b) for varying side lengths d.

rises because of the increased capacitive effect [6]. For a certain angular frequency ω, the phase velocity v_p is inversely proportional to the propagation constant β as follows:

$$v_p = \frac{\omega}{\beta} \tag{4.2}$$

The phase velocity of a wave would decrease in proportion to an increase in the propagation constant with height. If the blocks are quite high, field concentration is still another notable aspect of the suggested SWS. The electric field's intensity and concentration inside a unit of the SWS vary with block height, as shown in Figure 4.4. Because of the increased capacitive effect between blocks, the electric field is focused and intensified in the area between two face-to-face blocks as block height rises. This physical action might be employed to modify the amplitude of the field distribution on the horn aperture and increase the electric field intensity.

Because of their smaller top surface area, the triangular block units 1, 6, 10, and 13 have a lower capacitive coupling than the rectangular block. This contributes to a uniform phase distribution on the aperture by guaranteeing that the phase

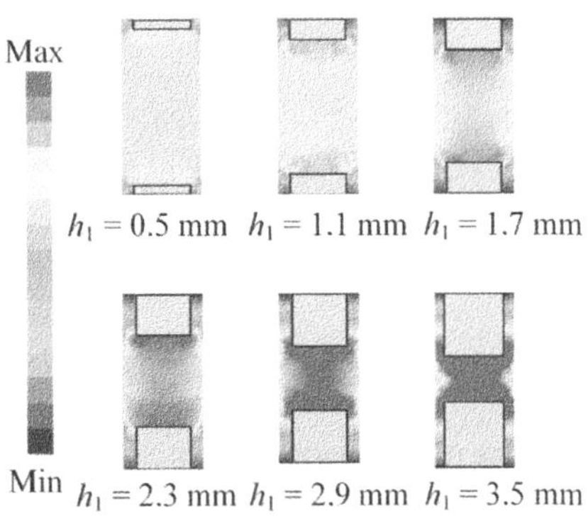

Figure 4.4 *E*-field magnitudes at various heights (at 12.5 GHz) in an SWS unit.

velocity close to the sidewall is quicker than the centerline. In order to improve impedance matching, the triangular block unit also seeks to lessen the discontinuity close to the sidewall.

4.1.2.3 Design of SWSs

As shown in Figure 4.1, the planned SWS is made up of a series of periodic metal blocks within the horn antenna that are symmetrically organized in both square and triangular configurations with regard to the horn antenna's center plane. Both the longitudinal and lateral periods, p and d, are set at 5 mm and 3.5 mm, respectively, for the block unit d.

To counteract the greater phase difference on the aperture caused by the horn's reduced longitudinal length, the wave traveling toward the aperture center should have a slower phase velocity than the wave traveling toward the aperture edges. Blocks in the midline ought to be taller than the blocks on either side of the aperture to achieve a more uniform phase distribution. The SWS-I design in Table 4.2 has blocks that are 3.3 mm high in the centerline and other blocks that are 2.3 mm high.

The phase distributions of the shortened horn loaded with SWS-I and the original horn are contrasted in Figure 4.5, which also shows the shorter horn without SWS. Figure 4.5c shows that SWS-I achieves a comparatively flat phase front. The gain of a horn antenna is also influenced by the field's magnitude distribution on the aperture. An energy confinement problem may arise because of the concentration of energy close to the aperture's center in Figure 4.5f. In order to get a more consistent amplitude and phase distribution on the aperture, the height of the blocks in the centerline is subsequently changed, producing SWS-II, which is shown in Table 4.3. The uniformity of the field strength across the aperture can be enhanced by gradually reducing the heights of blocks 1, 2, 3, and 4 along the centerline. This will allow more energy to flow to the aperture's side sections. SWS-II is capable of reducing the phase velocity even when the pin heights are nonuniform. This is evidenced by the observation that the guided wavelength in Figure 4.6b is shorter than that in Figure 4.5b at the same frequency. This conclusion is drawn

Table 4.2 Blocks' Height of SWS-I.

Block number	Height (mm)	Block number	Height (mm)	Block number	Height (mm)	Block number	Height (mm)
1	3.3	5	3.3	9	2.3	13	2.3
2	3.3	6	2.3	10	2.3	14	2.3
3	3.3	7	2.3	11	2.3	15	0
4	3.3	8	2.3	12	2.3	16	0

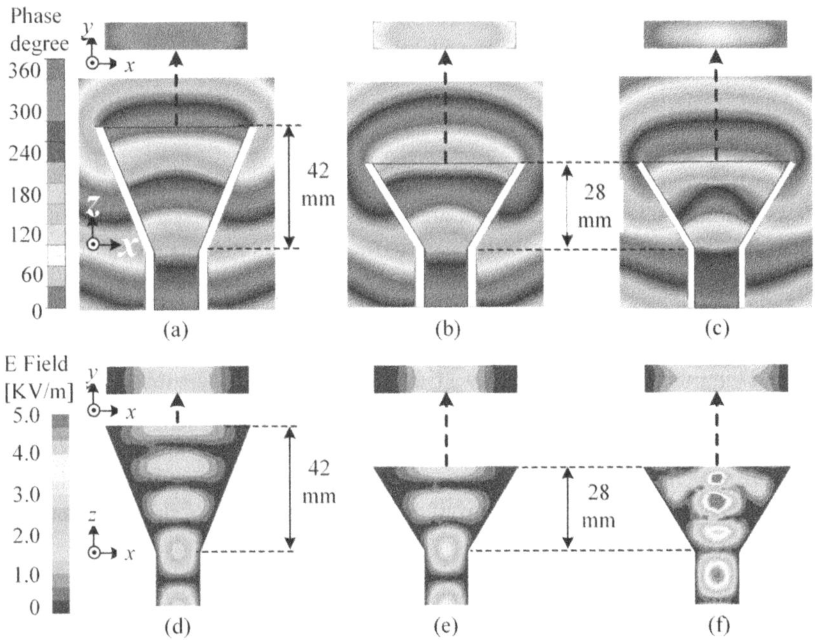

Figure 4.5 Illustration of the phase and magnitude distributions of the *E*-field at 12.5 GHz. (a) Phase distribution of a normal horn. (b) Phase distribution of a shortened horn without SWS. (c) Phase distribution of a shortened horn with SWS-I. (d) Magnitude distribution of a normal horn. (e) Magnitude distribution of a shortened horn without SWS. (f) Magnitude distribution of a shortened horn with SWS-I.

Table 4.3 Blocks' Height of SWS-II.

Block number	Height (mm)	Block number	Height (mm)	Block number	Height (mm)	Block number	Height (mm)
1	**2**	5	3.3	9	2.3	13	2.3
2	**2.3**	6	2.3	10	2.3	14	2.3
3	**1.6**	7	2.3	11	2.3	15	0
4	**0.4**	8	2.3	12	2.3	16	0

Bold means the changed height compared with the SWS-I in Table 4.2.

by comparing the phase distributions in Figure 4.5b (short horn without SWS) and Figure 4.6b (short horn with uneven-height SWS-II).

4.1.2.4 Impedance Match Improvement

The loaded SW blocks significantly affect the wave impedance of the flared waveguide inside the horn. This leads to an impedance mismatch between the feeding

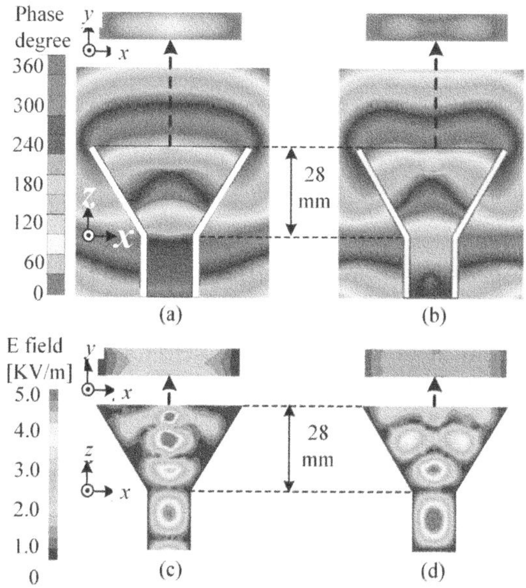

Figure 4.6 Phase and magnitude distributions of the *E*-field at 12.5 GHz. (a) Phase distribution of a shortened horn with SWS-I. (b) Phase distribution of a shortened horn with SWS-II. (c) Magnitude distribution of a shortened horn with SWS-I. (d) Magnitude distribution of a shortened horn with SWS-II.

waveguide and the flare region (near the horn's throat), as well as between the wave impedance of free space and the radiating aperture. Two stepped-height transition structures are developed close to the horn's throat and radiating aperture to enhance impedance matching.

The rectangular and flared waveguides of the horn antenna can be modeled as a transmission line with distributed series inductance L and shunt capacitance C, as illustrated in Figure 4.7. The radiation resistance Z_r of the horn aperture and the wave impedance Z_0 of empty space are linked in series with the antenna. The values of the SWS loaded in the throat and close to the aperture are determined by the height of the blocks and are shown in the model as ΔC_1 and ΔC_2, respectively. The wave impedance equation may be used to estimate the wave impedance Z_w at the radiating aperture.

$$
\begin{aligned}
Z_W &= \frac{k \cdot \eta}{\beta} \\
&= \frac{\omega \cdot \sqrt{\mu \cdot \varepsilon_{r\_eff}} \cdot \sqrt{\frac{\mu}{\varepsilon_{r\_eff}}}}{\sqrt{k^2 - k_c^2}} \\
&= \frac{\omega \cdot \mu}{\sqrt{\omega^2 \cdot \mu \cdot \varepsilon_{r\_eff} - \frac{\pi^2}{W_h{}^2}}}
\end{aligned}
\tag{4.3}
$$

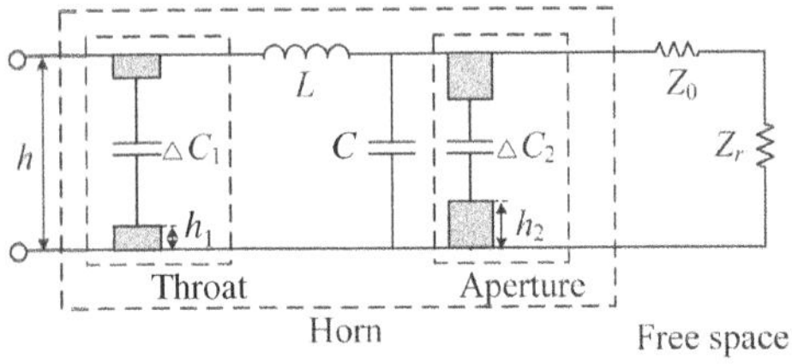

Figure 4.7 A simplified transmission line model of the proposed horn for impedance matching design.

where $k = \omega \cdot \sqrt{\mu \cdot \varepsilon_{r\_eff}}$ is the wave number, $\eta = \sqrt{\mu \cdot \varepsilon_{r\_eff}}$ is the wave impedance, and $\beta = \sqrt{k^2 - k_c^2}$ is the propagation constant at a specific frequency. The permeability μ, the relative effective dielectric constant ε_{r_eff}, and the aperture width W_h are all connected to the wave impedance Z_W. The relative effective permittivity ε_{r_eff} is influenced by the loaded SWS since the E-field is mostly centered between the higher and lower blocks. This is explained as [7].

$$\varepsilon_{r\_eff} = \varepsilon_0 \cdot \frac{h}{h - 2 \cdot h_1} \tag{4.4}$$

where ε_0 is the permittivity of air. The impedance of the waveguide at the radiating aperture decreases as the height of the blocks, h_1, increases because this causes a bigger effective relative dielectric constant. As a result, the radiating aperture and the open space have different impedances. The heights of blocks 9, 12, and 14 that are closest to the horn aperture are optimized to 1.6 mm in order to improve the impedance matching between the aperture and empty space.

However, as Figure 4.1b illustrates, two stepped transition blocks are added to improve the impedance matching between the feeding waveguide and the horn's flared section. The heights of these blocks are 0.4 and 1 mm. The blocks' arrangement has been changed to SWS-III, the evolved horn's last structure (Table 4.4). When Figure 4.8 plots the reflection coefficients of the shorter horn loaded with SWS-I, SWS-II, and SWS-III, it is evident that $|S_{11}|$ has significantly improved.

Table 4.4 Blocks' Height of SWS-III.

Block number	Height (mm)	Block number	Height (mm)	Block number	Height (mm)	Block number	Height (mm)
1	2	5	3.3	9	**1.6**	13	2.3
2	2.3	6	2.3	10	2.3	14	1.6
3	1.6	7	2.3	11	2.3	15	**1**
4	0.4	8	2.3	12	**1.6**	16	**0.4**

Bold means the changed height compared with the SWS-II in Table 4.3.

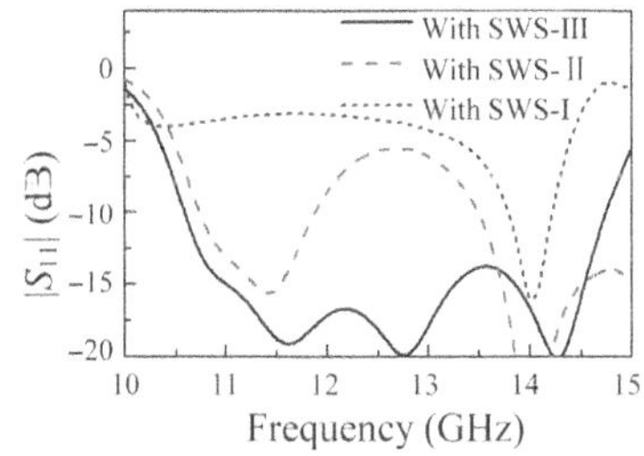

Figure 4.8 Simulated $|S_{11}|$s of the short horn loaded with SWS-I, SWS-II, and SWS-III.

4.1.3 Simulation and Measurement Results

An aluminum-based prototype of the designed antenna is made in order to validate the design idea. Figure 4.9a shows the photographs of the prototypes. Figure 4.9b shows the achieved gains and the measured and simulated $|S_{11}|$. Within the frequency band of 10.5–14.5 GHz, the created shorter horn loaded with SWS (Figure 4.1) shows an average gain that is 1.7 dB greater than the original horn (Figure 4.2a), which translates to a 32% fractional bandwidth. Figure 4.10 displays the observed radiation patterns in the E- and H-planes, demonstrating stability across the entire frequency band. These findings show that the horn antenna's maximum gain and reduced size have been accomplished at the same time.

One important metric for assessing aperture antenna performance is aperture efficiency. According to this definition, it is the ratio of the physical aperture area (A) to the effective aperture (A_e) as follows:

$$\eta_a = \frac{A_e}{A} \tag{4.5}$$

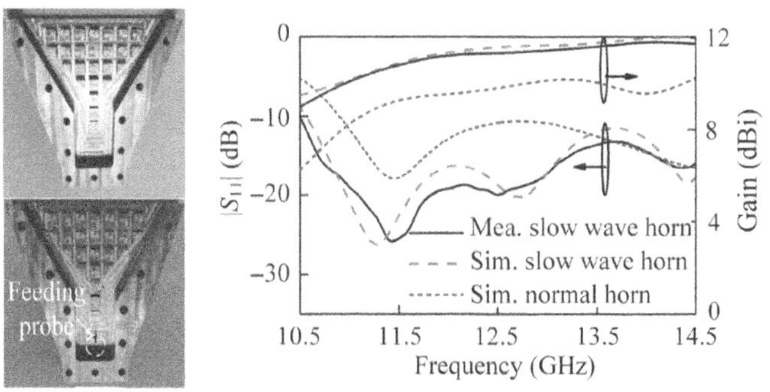

(a) (b)

Figure 4.9 (a) Images of the proposed horn loaded with SWS. (b) Gains and $|S_{11}|$s.

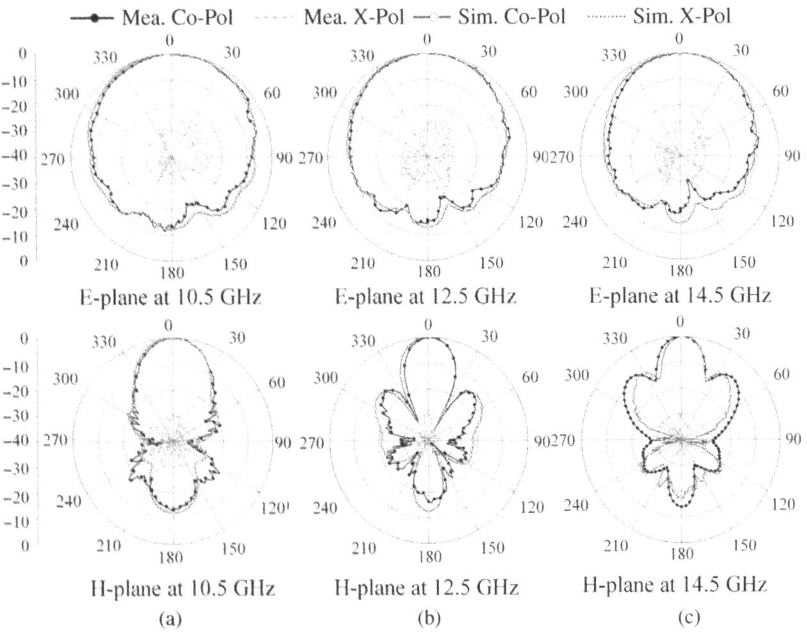

Figure 4.10 Experimental and simulated radiation patterns of the horn antenna loaded with SWSs at (a) 10.5, (b) 12.5, and (c) 14.5 GHz.

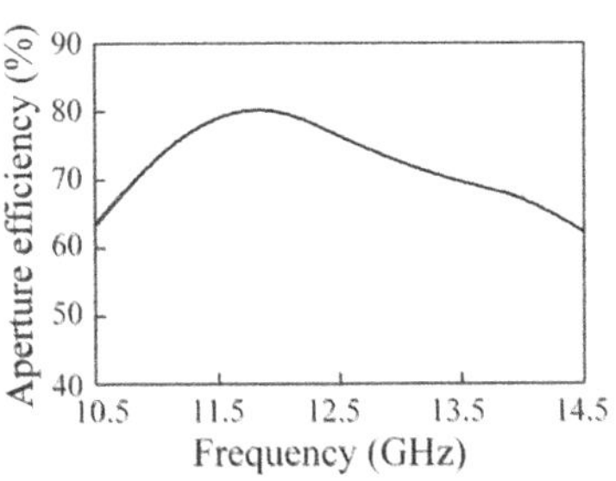

Figure 4.11 The aperture efficiency of the proposed horn plotted against frequency.

A_e is expressed as

$$A_e = \frac{\lambda^2}{4\pi} \cdot G \tag{4.6}$$

where G stands for the gain. Using the measured gains, Figure 4.11 displays the produced horn antenna's aperture efficiency. At 12 GHz, the aperture efficiency may increase to a maximum of 80.7%. Our design has a greater aperture efficiency than existing documented metal miniature horn antennas (Table 4.5).

Table 4.5 Comparison of Horn Antennas.

References	Frequency (GHz)	Aperture size ($\lambda_0 \times \lambda_0$) at center frequency[a)]	Gain (dBi)	Maximal aperture efficiency (%)	Adopted technology	Fabrication technology
[8]	5.5–7	8.53 × 4.27	21.2–22.5	38.87	Metamaterial lens	Only simulation
[9]	10–11.1	2.17 × 2.17	14.5–15.8	64.48	Bulk metamaterial	Six-layer substrate
[10]	10.2–11.1	5.9 × 4.9	22.5–23.5	60.58	Metal baffles	Metal processing
[11]	26–40	5.5 × 0.83	11.5–14.5	46.57	Slow-wave structures	Metal processing
This work	10.5–14.5	2.5 × 0.57	9.3–11.8	80.7	Slow-wave structures	Metal processing

a) The aperture size includes the thickness of sidewalks.

4.1.4 Conclusion

In this work, by adding periodic metal elements as embedded SWSs, we were able to reduce the longitudinal direction of an H-plane horn antenna by 33.3% in comparison to a standard horn. In addition, the aperture efficiency has been effectively increased, and the average gain is improved by 1.7 dB compared to the original design across the 10.5–14.5 GHz band. Furthermore, loading transitions at the feeding waveguide's throat and close to the radiating aperture enhance impedance matching. The design procedure shows that the proposed SWSs may simultaneously maximize aperture efficiency and minimize dimensions. There is potential for this little horn antenna to be used as a reflector feeder in radar and other radio systems.

4.2 Miniaturized H-Plane Horn Antenna in Longitudinal Direction Featuring −30 dB Sidelobes via Simple Blocks for Aperture Field Adjustment

4.2.1 Introduction

This research proposes a longitudinally downsized H-plane horn antenna. By spreading the field on the radiation aperture using rectangular blocks, the antenna seeks to reduce sidelobes. In comparison to an ideal horn, the design shortens the horn's longitudinal length by around 50%. Nevertheless, the electric forces on the aperture have nonuniform phase and nontapering magnitude distributions due to this length reduction, which results in large sidelobes and reduced gain. The suggested antenna packs four pairs of rectangular blocks close to the radiation aperture in order to solve these problems. While preserving a somewhat consistent phase profile on the aperture, these blocks disturb the fields and disperse the field magnitudes. Consequently, the antenna attains an average gain of 10.6 dBi across the 10.5–15 GHz frequency range, and the sidelobes are notably suppressed to below −30 dB. A prototype antenna is made and tested in order to validate the suggested antenna's performance. The simulation results show excellent consistency with the measured data. Compared to existing documented tiny antennas, the suggested H-plane horn antenna has two advantages: an ultra-low sidelobe level (SLL) of −30 dB and a straightforward shape that makes construction simple and inexpensive.

We present a longitudinally downsized H-plane horn antenna with ultra-low SLL and good gain in this article. This is accomplished by the suggested antenna by modifying the radiating aperture's phase and magnitude distributions. Four pairs of metal blocks are positioned close to the aperture in order to achieve these modifications. Relatively tapered and homogeneous distributions are the outcome of this magnitude and phase redistribution. Consequently, the gain is preserved but the sidelobes are notably reduced. The manufactured horn prototype has a fractional bandwidth of 35% and a broad impedance bandwidth between 10.5 and 15 GHz. The suggested horn's aperture measures around $3\lambda_0 \times 0.4\lambda_0$ (70 mm × 9.53 mm). Throughout this setup, the SLLs are less than −30 dB throughout the operational bandwidth, and the horn achieves an average gain of 10.6 dBi.

4.2.2 Antenna Design

Figure 4.12 shows the miniature H-plane horn antenna filled with four pairs of metal blocks near the center of the radiation aperture. The horn's aperture measures $3\lambda_0 \times 0.4\lambda_0$ (70 mm × 9.53 mm). On the horn's tapering H-plane walls, the symmetrical arrangement is applied to the four pairs of rectangular blocks. The horn's feeding waveguide's inner walls are made to the same size as the waveguide type WR-75. To excite the horn antenna, a coaxial probe is placed 6.4 mm from the short wall. Table 4.6 provides a detailed summary of the reduced H-plane horn antenna's design characteristics.

The ideal horn is the source of the miniature horn. Both the horn's length and the horn aperture's breadth must meet certain requirements to improve aperture efficiency. In accordance with the optimum horn theory, which guarantees that the phase variation from the aperture's center to its edge is below $3\pi/4$, the relationship between the horn's length and aperture width is assessed as follows:

$$R_h = \frac{a_1^2}{3\lambda} \tag{4.7}$$

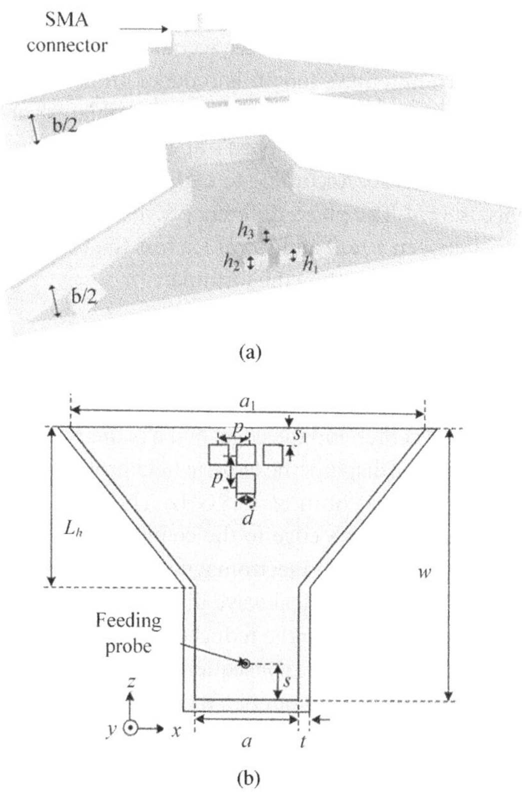

Figure 4.12 (a) 3-D structure. (b) Top view of the designed antenna.

Table 4.6 Structural Dimensions of the Proposed Antenna.

Parameters	Value (mm)	Parameters	Value (mm)	Parameters	Value (mm)
a	19.05	L_h	28	t	1
a_1	70	w	47.8	h_1	2
b	9.53	p	5	h_2	2
s	6.4	d	3.5	h_3	2
s_1	3.5	g	1.5	L_0	59

where λ represents the free-space wavelength, a_1 is the aperture width, and R_h is the flare radius. Thus, for a given aperture size, excellent aperture efficiency and gain might be achieved. According to formula (4.7), for an ideal horn, the length R_h is computed as 73 mm at 15 GHz, the operational band's highest frequency, with the aperture width fixed at 70 mm (3λ at 12.5 GHz). The ideal horn's length, L_0, is determined to be 59 mm after accounting for the waveguide's width.

A horn antenna's length is reduced from 59 to 28 mm or 52.9% less than the ideal horn length in order to achieve miniaturization. However, the phase inaccuracy on the shrunken horn's aperture becomes worse as a consequence of this length decrease. According to earlier research [11], the ideal horn's phase difference is $3\pi/4$. The phase difference across the aperture with the reduced horn may vary between $10\pi/9$ and $19\pi/12$ across the full frequency range (10.5–15 GHz), as described by the following formula:

$$\psi = \frac{\pi a_1(a_1 - a)}{4\lambda L_h} \tag{4.8}$$

where ψ is the phase difference of the electromagnetic wave between the horn aperture's center and the edge, and a is the rectangular waveguide's width.

Figure 4.12 displays the electric field profiles on the H-plane of the shorter horn and the optimal horn at 12.5 GHz. The reduced horn's greater phase variation from the aperture's edge to the center is emphasized by the electric field distributions. The ANSYS Electromagnetics Suite (HFSS) was used to model these distributions. The spherical wave divides as it propagates toward the horn aperture because the E-walls in the reduced horn are tilted more away from the z-axis (parallel to the y-axis). As a consequence, the E-fields in the radiating aperture's center have a phase that is about half a period ahead of the edge. A reduced directivity is a result of the aperture's increased phase inaccuracy.

The SLL of the reduced horn is shown against frequency in the H-plane in Figure 4.13. At 12.5 GHz, the SLL rises to −8 dB at an angle θ of around 40°. It is

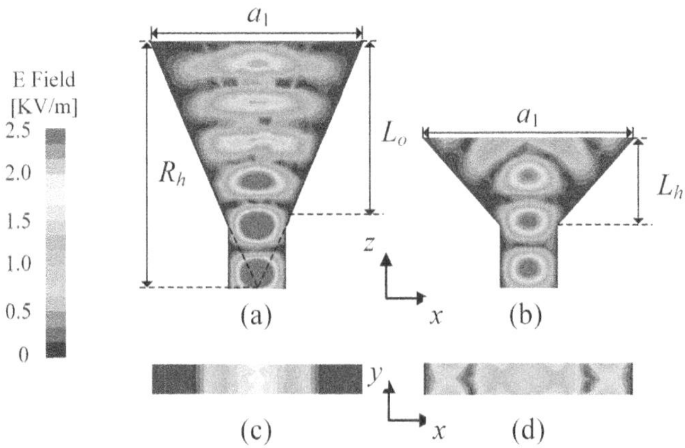

Figure 4.13 Electric field distribution of (a) the optimal horn and (b) the shortened horn and on the aperture of (c) the optimal horn and (d) the shortened horn at 12.5 GHz.

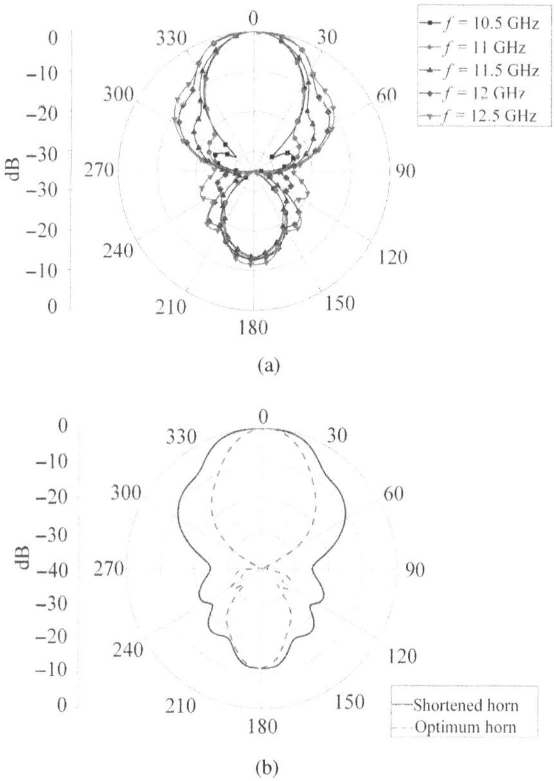

Figure 4.14 (a) H-plane radiation patterns of the short horn versus frequency. (b) Comparison of radiation patterns for the optimum horn and the shortened horn at 12.5 GHz.

clear that as the large-aperture horn's longitudinal length is decreased, the SLL rises noticeably in comparison to an ideal horn's SLL of around −32 dB.

Figure 4.14 displays the SLL of the reduced horn in the H-plane as a function of frequency. The SLL increases to −8 dB at an angle θ of around 40° at 12.5 GHz. In contrast to the ideal horn's SLL of around −32 dB, it is evident that the SLL increases substantially when the longitudinal length of the large-aperture horn is reduced.

4.2.3 Design Guideline and Working Principle

The following is a summary of the design guidelines for the suggested horn antenna: (1) Based on the ideal horn design criteria for a certain aperture size, calculate the horn length. (2) To accomplish downsizing, decrease the horn length while maintaining the same aperture size. (3) To achieve a tapered and uniform design for the magnitude and phase distributions, place metal blocks near the radiating aperture and maximize their dimensions.

The suggested miniature horn's operation involves loading metal blocks to disperse the field on the radiating aperture. A tapering magnitude distribution and a comparatively homogeneous phase distribution on the aperture are the goals of this redistribution. The transmission line model has a larger shunt capacitor as a consequence of the metal blocks' enhancement of the capacitance effect between the horn's top and bottom metal walls. By concentrating the electric field, this extra capacitor makes it possible to modify the field's strength. The extra capacitance also raises the effective dielectric constant ε_{r_eff} in comparison to the initial medium devoid of metal blocks [12], according to [6].

$$\varepsilon_{r-f\,eff} = \varepsilon_0 \cdot \frac{b}{b - 2 \cdot h} \tag{4.9}$$

The impact of the extra capacitors on the field distribution on the radiating aperture allows for the modification of the EM waves' phase and amplitude. This modification is accomplished by altering the loaded metal blocks' size characteristics.

The height of the metal block is the most critical dimension parameter as it has a big impact on the equivalent capacitance between face-to-face blocks. As shown in Figure 4.15, raising the metal block's height shortens the space between face-to-face blocks, which improves the equivalent capacitor. This allows for the modulation of the electric field's magnitude since it concentrates on the area between face-to-face blocks. Furthermore, a delay in the EM waves' passage through the face-to-face blocks is introduced by the increase in the equivalent capacitor, leading to an extra phase shift and permitting phase modifications.

Full-wave modeling software was used to generate miniature horns with different metal block heights in order to confirm the impact of metal blocks on magnitude and phase distribution changes. Figure 4.16 shows how the height of the metal block affects the electric field distribution on the aperture at 15 GHz. It is

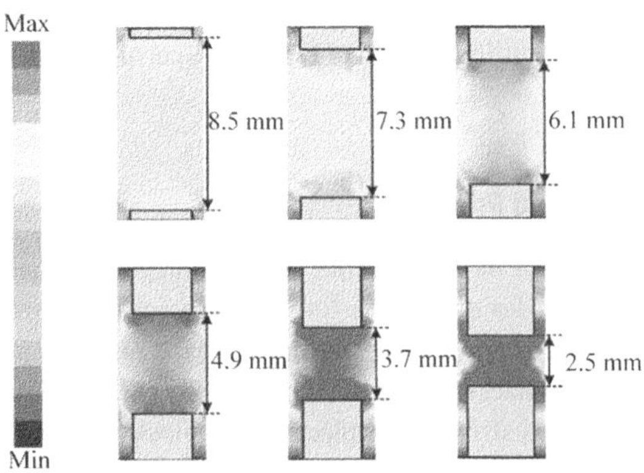

Figure 4.15 Impact of the separation between two metal blocks on the field profile.

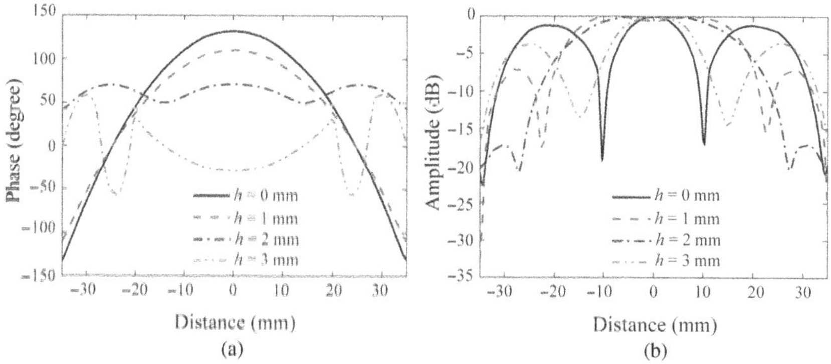

Figure 4.16 Simulated E-field distributions of short horn with blocks of different heights at 15 GHz. (a) Phase distribution over the aperture of the horn in the H-plane. (b) Magnitude distribution over the aperture of the horn in the H-plane.

evident from Figure 4.16a that the enlarged equivalent capacitor causes the phase at the aperture's center to delay as block height rises. The electric field's phase distribution on the aperture stabilizes at a height of 2 mm for the loaded blocks. Additionally, Figure 4.16b indicates that the electric field becomes more concentrated toward the center of the radiating aperture as the metal block's height increases. The middle portion of the horn is where most of the electric field energy on the aperture is focused for the loaded blocks that are 3 mm high. To obtain the aperture's phase distribution to be almost uniform, the metal block heights (h_1, h_2, and h_3) are chosen as 2 mm.

The amplitude and phase distributions of the E-fields for the optimal horn and the truncated H-plane horn with and without metal blocks are shown graphically in Figure 4.17. The E-field's magnitude and phase distribution may be changed by applying metal blocks to the inner walls of the H-planes close to the aperture. As shown in Figure 4.18, this modification lowers the high SLL by producing a more even phase distribution and a smoothly varying amplitude distribution on the downsized H-plane horn's aperture. Additionally, the shorter horn with metal blocks exhibits a radiation pattern that is quite similar to the ideal horn.

4.2.4 Simulation and Measurement Results

An aluminum alloy prototype is made in order to confirm the design procedure and the simulated outcomes of the miniature H-plane horn antenna. The photographs of the fake structure are shown in Figure 4.19a. An Agilent Vector Network Analyzer (E8363C) is used to assess the prototype's S_{11} parameter magnitude, and a far-field radiation pattern measurement device is used to get the radiation patterns. Figure 4.19b displays the observed and simulated reflection

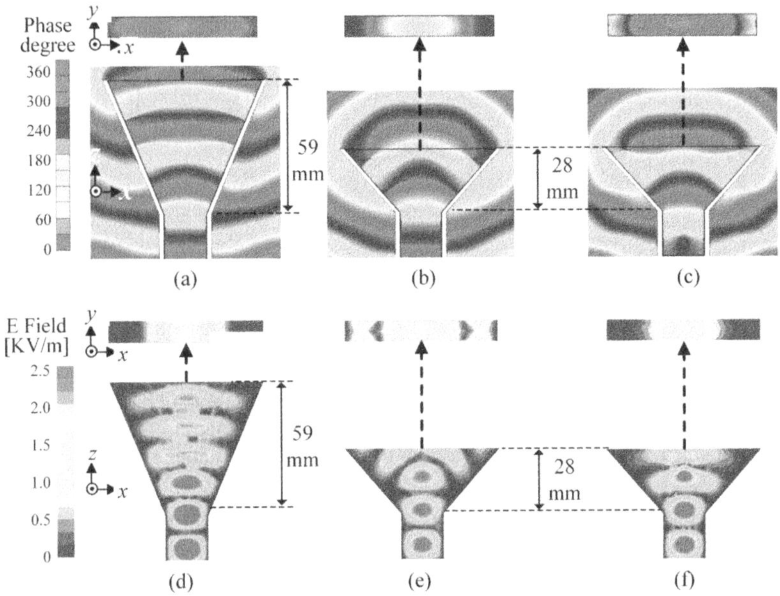

Figure 4.17 Simulated E-field distributions at the H-plane of the optimum horn, the short horn without metal blocks, and the proposed short horn with metal blocks at 12.5 GHz, including (a)–(c) phase distribution and (d)–(f) magnitude distribution.

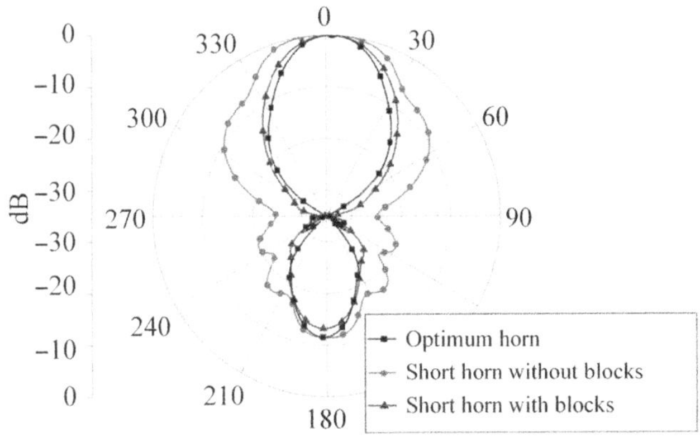

Figure 4.18 H-plane radiation patterns at 12.5 GHz simulated for the optimal horn and the short horn, with and without metal blocks.

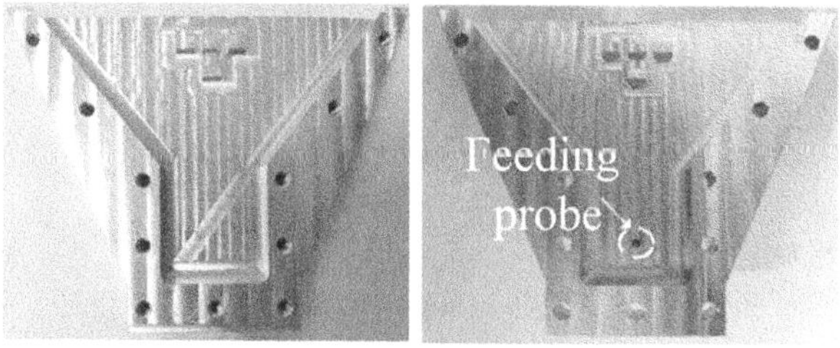

(a)

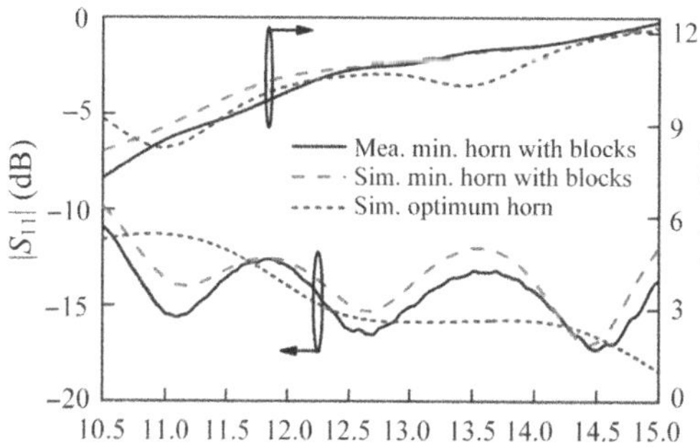

(b)

Figure 4.19 (a) Images of the created prototype. (b) Measured and simulated gains and $|S_{11}|$s of the proposed horn compared with the optimum horn as a function of frequency.

coefficient and realized gain findings. A 35% bandwidth is covered by the S_{11} parameter, which has a magnitude of less than −10 dB between 10.5 and 15 GHz. Furthermore, there is excellent consistency between the simulation findings in HFSS and the actual measurements, since the realized gain nearly resembles that of the ideal horn.

The observed and simulated normalized radiation patterns at various frequencies in the H-plane are shown in Figure 4.20. The observed cross-polarization levels are around −30 dB, but the simulated values are less than −35 dB at most angles. Throughout the whole frequency range, the SLLs in the observed H-plane patterns are all less than −30 dB. Figure 4.21 displays the radiation patterns of the horn design from prior research for comparison purposes. It is clear that the present design's SLLs are much lower than those of the horn design from the earlier research. This work requires only four pairs of metal blocks, employing the mode perturbation approach to modify the field distribution, while the prior research depended on the SW effect obtained by 16 pairs of blocks. As a result, the design and manufacturing procedures are simpler than the SW horn.

This design not only has a smaller physical size but also produces a tapered amplitude distribution over the aperture, which significantly lowers the SLLs in

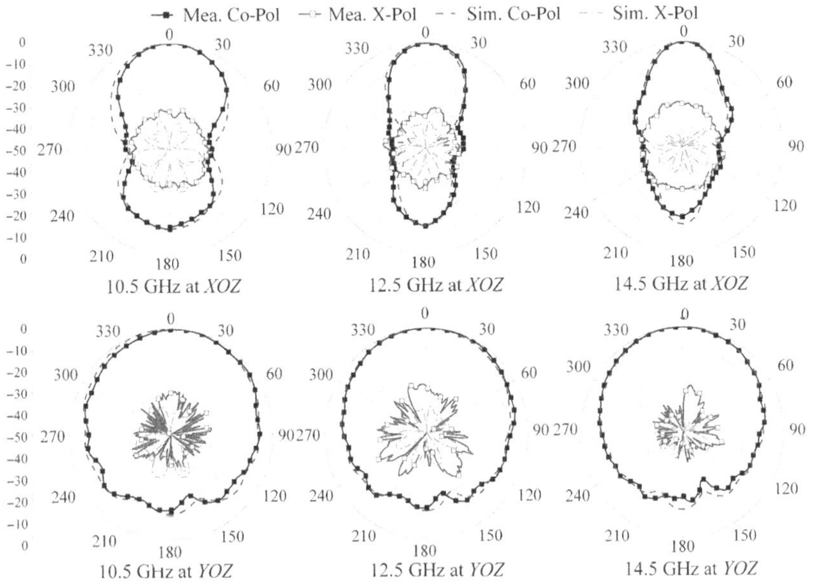

Figure 4.20 Measured and simulated radiation patterns of the proposed horn antenna across various frequencies.

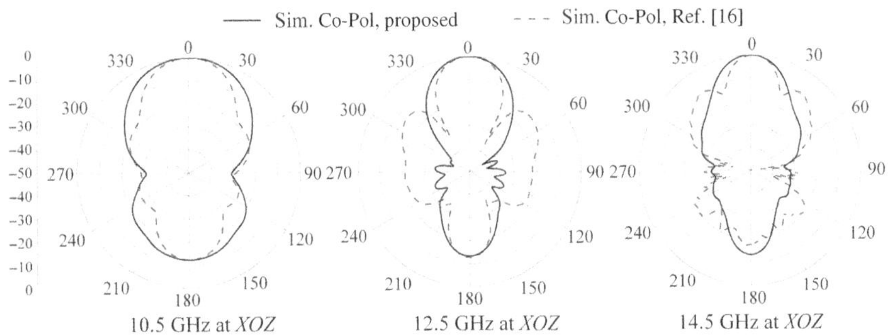

Figure 4.21 Radiation patterns of the proposed horn antenna. Source: [16]/IEEE.

the radiation pattern. On the other hand, this element is absent from the design in [15], which results in greater sidelobes. Figure 4.19 displays the observed and predicted radiation patterns in the E-plane, the primary beam of it is shown to align well, and its beamwidth is greater than that in the H-planes. At every tested frequency, the recorded cross-polarization values stay above −20 dB. These findings verify that the longitudinally shortened H-plane horn antenna has low SLLs and strong gain performance across a broad bandwidth after the loading of the metal blocks. Comparing the suggested horn's performance to other pertinent works,

Table 4.7 Dimensions of the Designed Antenna.

References	Adopted structure	Operating frequency (GHz)	Horn length (λ_0)	Aperture size ($\lambda_0 \times \lambda_0$)	Gain (dBi)	Aperture efficiency (%)	SLL (dB)
[13]	Dielectric lens	27	1.98	2.26 × 0.26	9.7	21.8	−10
[14]	Gap SIW	34	4.4	2.3 × 0.17	10.3	31.7	−14
[15]	SWS	12.5	1.17	2.5 × 0.57	11.3	76.2	−14
This work	Metal block	12.5	1.17	3 × 0.4	10.6	63.4	−30

Table 4.7 shows comparatively high gains and low SLLs. The observed findings closely resemble the predicted outcomes shown in Figures 4.18 and 4.19. Errors in machining and assembly are to blame for the minor difference between measured and simulated results. High-precision processing can reduce deviations because the height of the metal blocks, in particular, is a crucial parameter that influences the magnitude and phase distribution on the aperture.

4.2.5 Conclusion

This work successfully designed, built, and measured a miniature H-plane horn antenna with a longitudinal decrease of 52.9%. By placing rectangular blocks close to the radiating aperture, the antenna design used the field redistribution technique. This method made it possible to obtain a tapering magnitude distribution from the center to the sides as well as a uniform phase distribution of the electric fields on the aperture. Over a broad 35% bandwidth, the suggested horn antenna maintained a respectably low SLL of −30 dB and showed outstanding radiation characteristics. In the 10.5–15 GHz frequency band, it produced an average gain of 10.6 dBi.

4.3 Compact Slow-Wave SIW H-Plane Horn Antenna with Increased Gain for Vehicular Millimeter-Wave Communication

4.3.1 Introduction

Horn antennas can be used in the communication scenario shown in Figure 4.22, due to its attractive merits such as simple structure, high gain, wide bandwidth and high power capacity. Traditional metallic horn antennas always suffer from

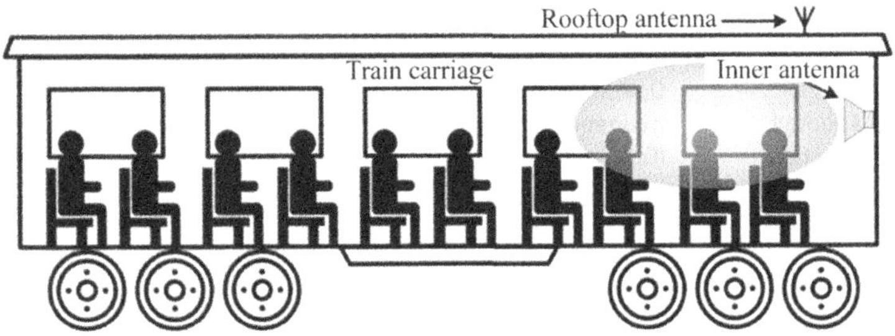

Figure 4.22 Communication scenario of terminal users in a train.

some obvious disadvantages of bulky size, heavy weight, high cost and difficulty of integration. With the invention of the substrate integrated waveguide (SIW), horn antennas can be implemented into planar structures with the convenience of planar integration, light weight and low cost. Generally, horn antennas are designed in accordance with the optimum horn criterion to get the maximum gain with a certain aperture size. Given the fact that the aperture size of a horn antenna determines its highest achievable directivity, longitudinally miniaturized SIW horn antenna without gain decrease is a competitive solution of the RF front-end design for space-limited applications. Straightforwardly reducing the longitudinal length of an optimum horn antenna will result in gain decrease, as the phase difference between the center and edge of the aperture is enlarged.

This letter suggests loading the SWS in the form of metalized blind via-holes to create an SIW H-plane horn antenna with longitudinally reduced dimensions and enhanced gain. The 28 GHz band (27.5–28.35 GHz) designated by the Federal Communications Commission for 5G purposes is covered by the planned SW-SIW horn. The longitudinal length of the SW-SIW horn is 53.83% less than the ideal SIW horn. Furthermore, the recommended SW-SIW horn improves its gain by 1.1 and 2.7 dB on average from 27.5 to 28.5 GHz when compared to the ideal SIW horn and the merely reduced SIW horn without SWS. Measurements and manufacturing of a prototype of the suggested SW-SIW H-plane horn antenna show good agreement with the calculations.

4.3.2 Antenna Design and Analysis

This section describes the design and analysis of the longitudinally downsized SIW H-plane horn antenna loaded with the SWS in the form of metalized blind via-holes. First, rectangular patches are loaded in front of the aperture to boost the bandwidth of an ideal SIW H-plane horn antenna. Then, as an intermediary step, the longitudinal length of this SIW horn with loaded patches is simply shortened. Crucially, the phase distribution on the horn aperture is calibrated to be

Table 4.8 Dimensions of the Designed Antenna Units: Millimeter.

Variable	A_h	p	d_1	d_2
Value	24	0.8	0.5	0.42
Variable	h_1	h_1	S_1	S_2
Value	4	0.508	1	0.3
Variable	W_s	W_p	L_h	L_s
Value	4	1.7	13.68	14.4
Variable	L_p	L_1	L_2	L_3
Value	6	2.65	1.15	2.22
Variable	W_1	W_2	W_3	L_4
Value	0.3	1.1	2	0.8
Variable	L_5	L_6	g_1	g_2
Value	1.6	4	0.15	0.45

somewhat uniform, and the SWS in the form of metalized blind via-holes is loaded within the horn to lower the phase velocity around the wide wall's centerline. The SW-SIW horn may be made smaller longitudinally without compromising gain, and its gain can be raised above that of the ideal horn. The parameters of the suggested SW-SIW H-plane horn are listed in Table 4.8, and its shape is shown in Figure 4.23a. To facilitate the use of the metalized blind via-holes, two Rogers 4350B substrates with a dielectric constant of 3.66 and a loss tangent of 0.0037 are used. Four components make up the suggested antenna: the SWS in the form of metalized blind via-holes inside the flare region, the impedance transformer in the form of metal patches loaded in front of the horn aperture, the flare region of the SIW H-plane horn, and the transition from SIW to grounded coplanar waveguide (GCPW). An ideal SIW H-plane horn with longitudinal length $L\_(h)' = 29.63$ mm serves as the model for the design. Then, the aperture width remains constant, while the longitudinal length is simply shortened to 13.68 mm. Lastly, to compensate for the gain loss, the SWS is injected into the flare zone through blind via-holes.

4.3.2.1 Original Optimum SIW H-Plane Horn

The optimum horn criteria are used to estimate the aperture width Ah and flare radius R_h' of an optimal SIW H-plane horn antenna [17]:

$$R_h' = \frac{A_h^2}{3\lambda} \tag{4.10}$$

The geometric relationship between the flare radius R_h', aperture width A_h, feeding SIW length W_s, and horn length L_h' may be used to get the horn length L_h'.

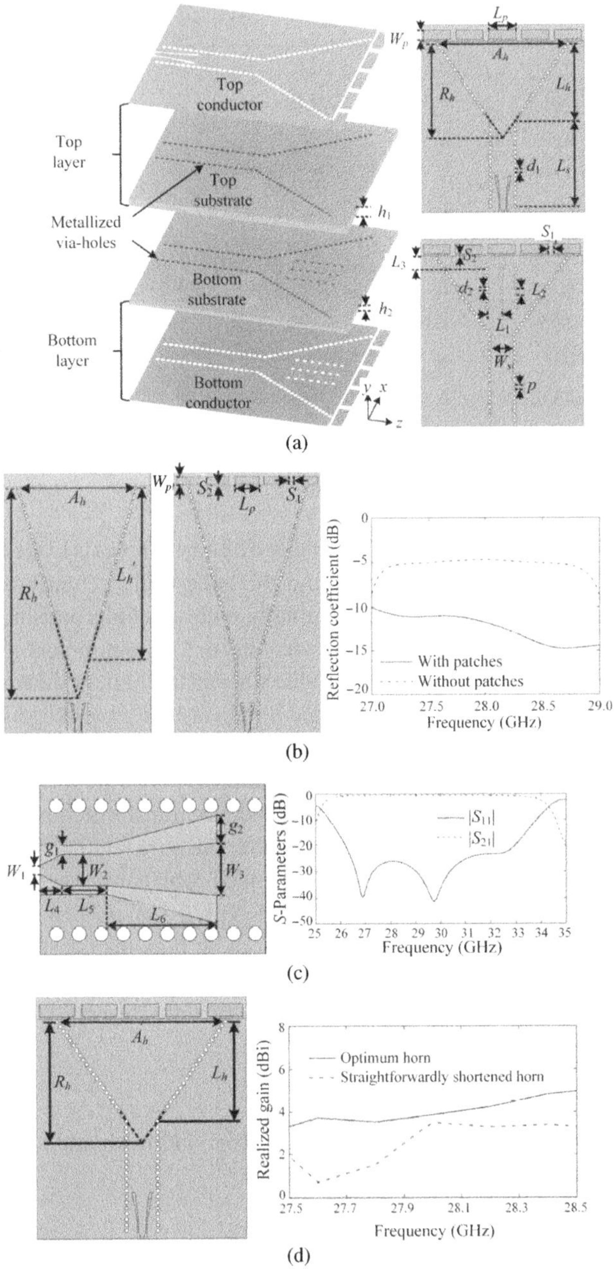

Figure 4.23 Geometrical configuration of SIW horns. (a) SW-SIW horn. (b) Optimum horn and reflection coefficient. (c) Transition and S-parameters. (d) Straightforwardly shortened horn and gain.

This connection is shown by the equation:

$$L'_h = R'_h - \frac{W_s R'_h}{A_h} \tag{4.11}$$

Figure 4.23b displays the optimal SIW horn arrangement. In order to achieve a cutoff frequency of 19.5 GHz, which is below the operating frequency's lower limit, the feeding SIWs' width is 4 mm. In accordance with SIW design guidelines, the via-holes' period (p) and diameter (d_1) are 0.8 mm and 0.5 mm, respectively [18]. A_h, the aperture width, is fixed at 24 mm. According to Eqs (4.10) and (4.11), the horn flare radius R'_h should not be less than 34.71 mm, and the horn length should not be less than 29.12 mm. In light of the metalized via-holes' period ($p = 0.8$ mm), the horn length L'_h is 29.63 mm.

As the substrate thickness is much less than the wavelength, there is a significant impedance mismatch between the SIW horn aperture and the open space [16, 19]. As shown in Figure 4.23b, to achieve a smooth impedance transition from the horn aperture to the open space, five pairs of double-sided patches, each measuring $L_p \times W_p$, are symmetrically loaded on the substrate's top and bottom surfaces ahead of the horn aperture.

Figure 4.23b shows the computed reflection coefficients of the ideal SIW horn with and without patches. In the whole 28 GHz band, the reflection coefficient of the ideal horn with patches is better than −10 dB, demonstrating that loaded patches may effectively enhance impedance matching.

The purpose of the GCPW-SIW transition is to enable measurement feasibility. Table 4.8 lists the transition's dimensions, whereas Figure 4.23c displays its shape. The transition is made up of two tapered slots, a 50 Ω GCPW, and a tapered GCPW that is coupled to a coaxial connection. Figure 4.23c depicts the transition's simulated S-parameters. In the whole 28 GHz range, the insertion loss is better than 0.4 dB, and the reflection coefficient is better than −20 dB between 26.35 and 33.11 GHz.

4.3.2.2 Straightforwardly Shortened SIW Horn

By simply decreasing its longitudinal length, a reduced SIW H-plane horn with an unchanged aperture width is depicted in Figure 4.23d. In comparison to the ideal horn illustrated in Figure 4.23b, the downsized SIW horn's longitudinal length is 13.68 mm, which represents a 53.83% decrease. The shrunken SIW horn's other dimensions, with the exception of longitudinal length, are identical to those of the ideal SIW horn antenna.

The wave front at the wide wall's center comes before the wave front at the edge of the flare area. It is a well-known fact that the simple reduction of the longitudinal length increases the phase difference between the horn's radiating aperture's edge and center. Ansys HFSS and CST Studio Suite are used to model the amplitude and phase distribution of the optimal and simply reduced horns, respectively.

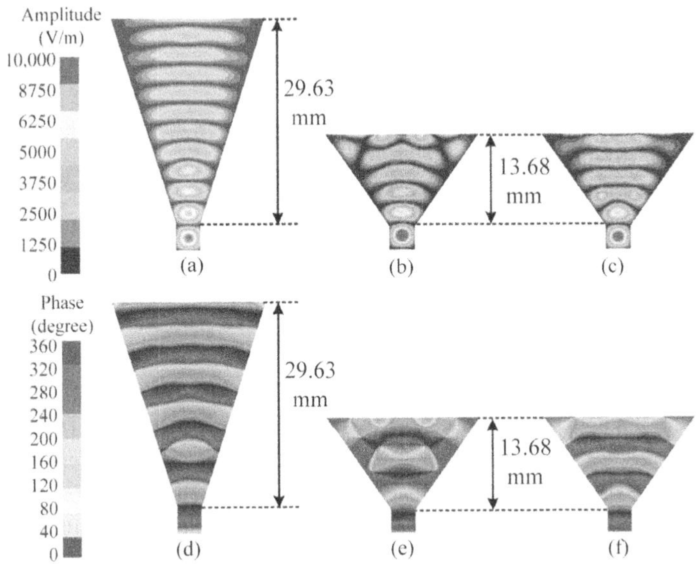

Figure 4.24 Phase and magnitude distributions of E-field at 28 GHz. (a) Magnitude of the optimum horn. (b) Magnitude of the straightforwardly shortened horn. (c) Magnitude of the SW-SIW horn. (d) Phase of the optimum horn. (e) Phase of the straightforwardly shortened horn. (f) Phase of the SW-SIW horn.

In contrast to the optimal horn, the phase difference of the simply reduced horn is increased, as shown in Figure 4.24. Figure 4.23d shows a comparison of the gains between the original optimal horn and the horn that was simply reduced. Between 27.5 and 28.5 GHz, the ideal SIW horn's average gain is 4.1 dBi, but the simply reduced horn's average gain is 2.5 dBi.

4.3.2.3 SIW H-Plane Horn Loaded with Slow-Wave Structure

An SWS in the form of metalized blind via-holes is included in the flare zone to increase the gain of the simply reduced horn antenna. Figure 4.23a shows how the SWS is configured. The via-hole h_2 is 0.508 mm high, which is the same thickness as the bottom substrate. The via-hole's height is limited by the substrate thickness that is available, as specified in the supplier's catalog. A higher SW via-hole produces a more noticeable SW effect when the SIW horn's overall thickness h is constant, but it also makes impedance matching more difficult [6, 20]. In order to balance the SW effect and the reflection coefficient, the height of the blind via-hole h_2 is fixed at 0.508 mm in this design.

In order to overcome the drop in gain caused by the decreased longitudinal length, it is essential to use the SWS to generate a somewhat uniform phase distribution on the horn aperture by reaching an adequate phase velocity along the wide wall's centerline. The phase velocity v_p can be determined using the following equation:

$$v_p = \frac{\omega}{\beta} \tag{4.12}$$

where ω is the angular frequency, and β is the propagation constant.

To show the link between the propagation constant and the total height (h_2) and spacing (d_2) of the substrates, Figure 4.24a simulates and displays the dispersion diagram of a unit cell of the SWS. The SW effect is strengthened as a result of the increased capacitance created between the metalized blind via-holes and the top ground layer, as shown by the observation that the propagation constant rises as h_2 and d_2 grow.

The electric field magnitude and phase distributions within the SW-SIW horn and the simply shorter horn are shown in Figure 4.24. It is discovered that the phase velocity slows down near the wide wall's middle line. Furthermore, it is seen that the electromagnetic wave in the plainly shortened horn is cylindrical, but the wave in the SW-SIW horn is comparatively flat. Additionally, it seems that the phase distribution on the SW-SIW horn's aperture is quite uniform.

Figure 4.25b shows the simulated reflection coefficients of these antennas, which include the SW-SIW horn, the optimal SIW horn, and the simply reduced SIW horn. Over the whole 28 GHz range, the reflection coefficients are better than −10 dB.

The simulated normalized H-plane radiation patterns of these three antennas at various frequencies are shown in Figure 4.25c. Due to the higher phase difference at the aperture caused by the lower longitudinal dimension, the H-plane patterns of the merely shortened horn are deformed when compared to the ideal horn, especially at 28.35 GHz, where the greatest radiation is no longer in the direction $\theta = 0°$.

4.3.3 Simulation and Measurement Results

Figure 4.26a illustrates the fabrication of a prototype of the suggested SW-SIW H-plane horn antenna utilizing the printed circuit board (PCB) manufacturing technique on two Rogers 4350B substrates.

The Agilent PNA E8363C Vector Network Analyzer is used to measure the reflection coefficient of the micro-SW-SIW horn. For 5G mm-wave applications, the observed reflection coefficients, which are depicted in Figure 4.26a, include the whole 28-GHz spectrum. Figure 4.26b displays the results of monitoring the radiation patterns and realized gain of the miniaturized SW-SIW horn in a microwave chamber. They show that the observed and anticipated radiation patterns agree pretty well.

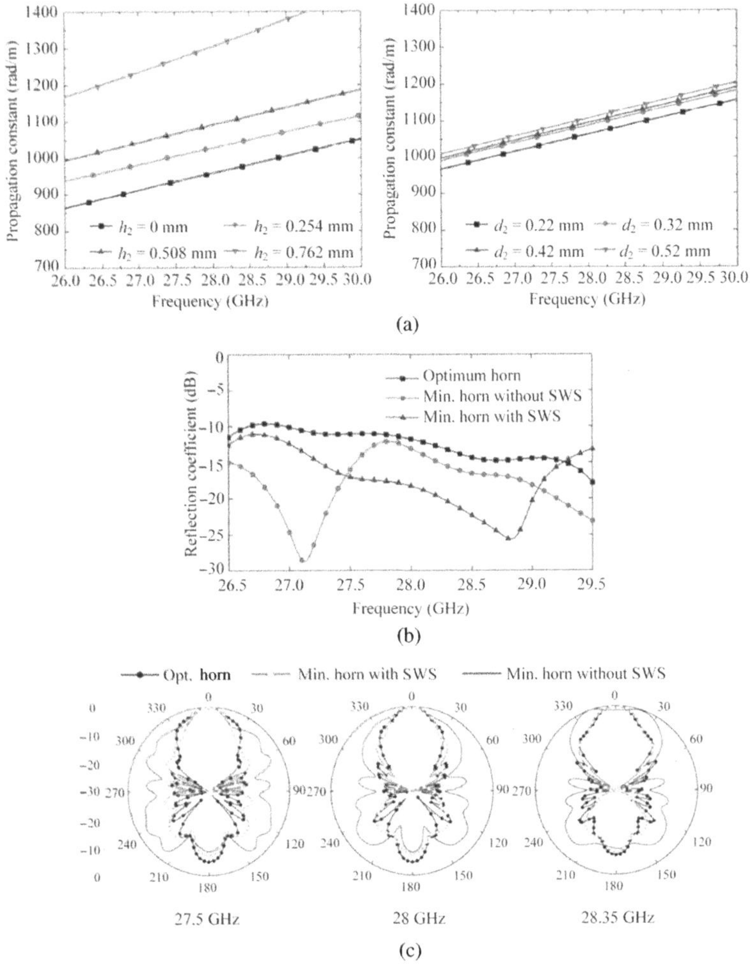

Figure 4.25 Simulated results of SIW horns. (a) Dispersion diagrams. (b) Simulated reflection coefficients. (c) Simulated radiation patterns.

Figure 4.26c illustrates the simulated and measured gains of the three antennas indicated above: the ultimate SIW horn, the simply shorter SIW horn, and the SW-SIW horn. The ideal SIW horn has a gain of 3.5–5.1 dBi, whereas the merely shorter SIW horn has a gain of 0.7–3.5 dBi throughout the working range, which is, on average, 1.6 dB less than the optimal SIW horn. The feasibility of the dimension reduction and gain enhancement of a SIW horn antenna enabled by SWS in

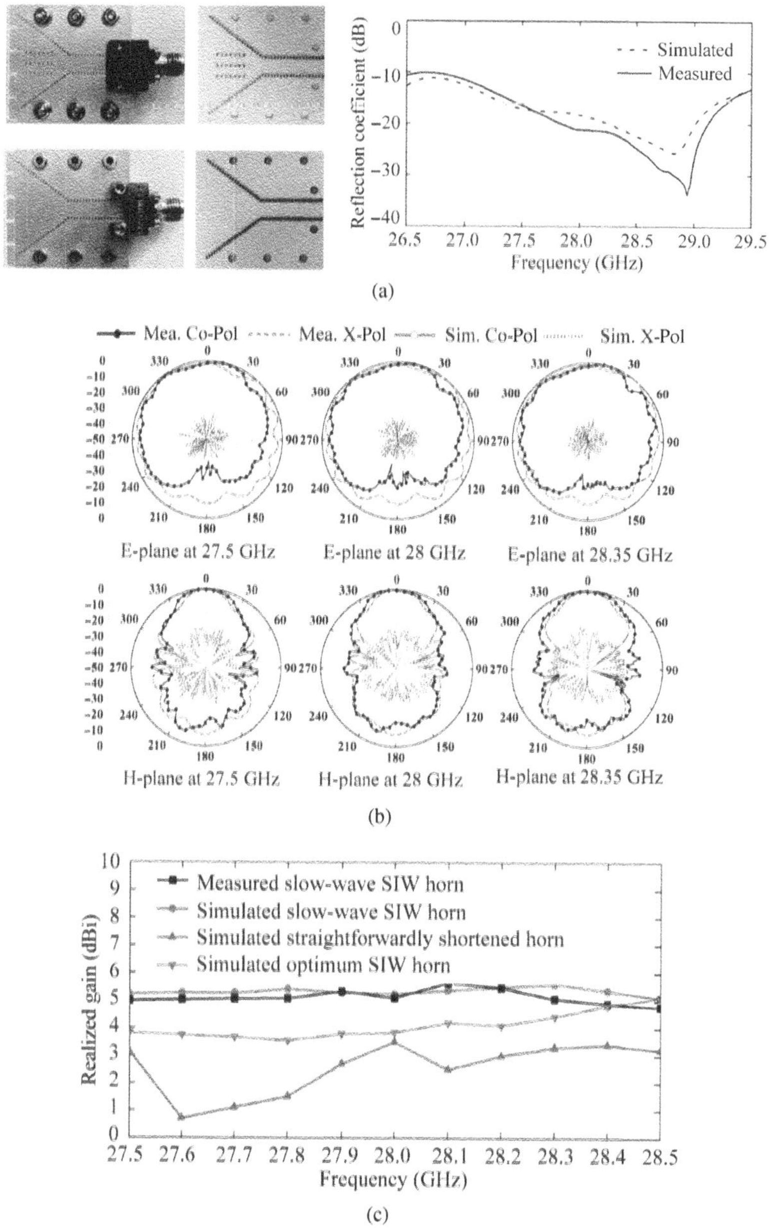

Figure 4.26 Comparison between simulated and measured results. (a) Prototype of the SW-SIW horn and measured reflection coefficient. (b) Simulated and measured radiation patterns. (c) Simulated and measured gains.

Table 4.9 Dimensions of Horn Antennas.

References	Type	Frequency (GHz)	Size ($\lambda \times \lambda \times \lambda$)	Gain (dBi)
[21]	SIW horn	30.0	2.61 × 0.17 × 3.46	5.5–10
[22]	SIW horn	16.6	1.80 × 0.08 × 3.26	4.0–5.5
[23]	SIW horn	35.5	1.89 × 0.18 × 3.47	8.5–9.3
[14]	SIW horn	35.0	2.30 × 0.17 × 4.40	5.4–6.8
[19]	SIW horn	20.0	1.59 × 0.33 × 3.90	7.9–9.5
Proposed	SIW horn	28.0	2.24 × 0.14 × 2.78	5.1–5.5

the form of metalized blind via-holes is demonstrated by the realized gain of the proposed SW-SIW horn antenna, which is above 5 dBi and improved by 1.1 dB and 2.7 dB, respectively, compared with the optimal SIW horn and the simply shortened SIW horn antenna. The projected results and the measured gain of the proposed small SW-SIW horn accord fairly well.

The suggested SW-SIW horn antenna's small size is made possible by its SWSs, as shown by a comparison of it with other SIW horns in Table 4.9.

4.3.4 Conclusion

By adding an SWS in the form of metalized blind via-holes within the horn's flare zone, this research suggests a miniature SW-SIW H-plane horn antenna with higher gain. In order to compensate for the increased phase difference on the horn aperture brought on by the horn's decreased longitudinal length, the SWS's geometries are tuned to provide a lower phase velocity around the wide wall's centerline. The Federal Communications Commission has designated the 28 GHz band (27.5–28.35 GHz) for 5G mm-wave applications, which is where the proposed miniature SW-SIW horn antenna works. Over 27.5–28.5 GHz, the SW-SIW horn's gain is enhanced by an average of 1.1 dB when compared to the optimal horn and by 2.7 dB when compared to a miniature horn devoid of SWSs. Additionally, without compromising radiation performance, the SIW H-plane horn antenna's longitudinal length is 53.83% shorter than that of the ideal SIW horn. Simulations, fabrications, and measurements have been conducted on the suggested SW-SIW horn. The viability of the dimension reduction and gain enhancement of a SIW horn antenna enabled by an SWS in the form of metalized blind via-holes is confirmed by the acceptable agreement between the simulated and measured findings.

4.4 Circularly Polarized Horn Antenna with Miniaturized Longitudinal Dimension for Vehicular Satellite Communications

4.4.1 Introduction

In satellite communications, circularly polarized (CP) antennas are essential for resolving issues with polarization mismatch in depolarizing media. Compared to sub-6GHz, high-speed satellite communications are more susceptible to transmission losses since they often operate at Ku-band and above. For direct connection with satellites, such as for customers in trains, Ku-band penetration losses are too large. As shown in Figure 4.27, the suggested approach entails positioning a directional indoor Wi-Fi antenna in the train compartment to provide high-density coverage and a small horn antenna on the train's roof to connect with satellites.

This research suggests using metal blocks and antipodal tapered slots (ATSs) to construct a wideband CP horn antenna with a reduced longitudinal length, resulting in a compact CP horn antenna without the need for extra polarizers or bigger antenna profiles. In order to obtain CP radiation, the ATSs are cut on the horn's large walls to create cross-polarization with approximately equal amplitude and quadrature phase in contrast to the horn's original linear polarization. The longitudinal length of the CP horn is reduced in order to make it smaller, which disrupts the two linear polarizations and impairs CP function. Three pairs of metal blocks have been installed next to the ATSs to address this, providing extra capacitances to balance orthogonal polarizations on the horn antenna aperture with a shorter longitudinal profile. At the central operating frequency, the suggested design achieves a downsized longitudinal length of $1.36\lambda_0$, which is a 33.3% decrease from the original ATSs' CP horn. By inserting metal blocks at the horn antenna's throat, the impedance matching is enhanced. Additionally, for both impedance and axial ratio (AR) bandwidths, the suggested antenna achieves an operational bandwidth of 32.8% between 11.35 and 15.8 GHz. Additionally, a

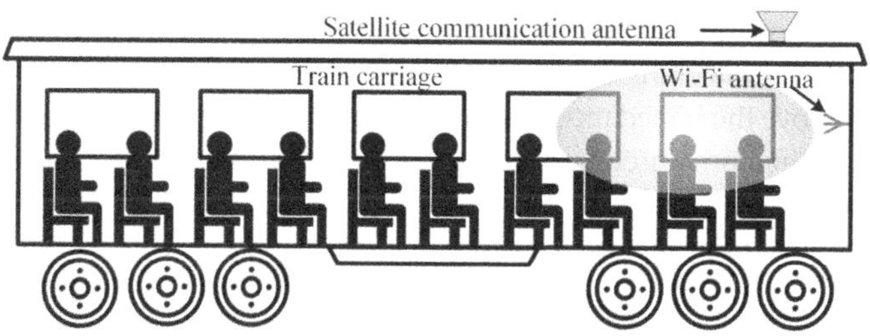

Figure 4.27 Satellite communications solution for carriage coverage.

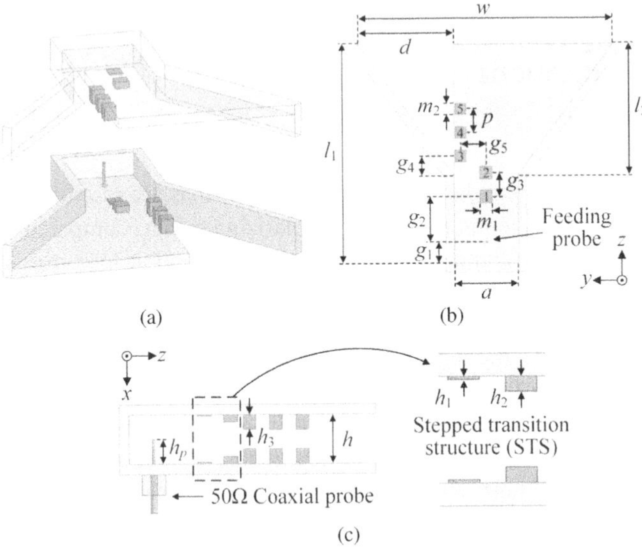

Figure 4.28 Configurations of the proposed CP horn antenna. (a) 3-D view. (b) Top view of the lower structure. (c) Cross-sectional view.

low-cost manufacturing procedure may be used to create the suggested design due to its straightforward structure.

4.4.2 Antenna Geometry

With tapered slots cut with artificial transmission surfaces and loaded with metal blocks, the suggested H-plane horn antenna designs are shown in Figure 4.28. The metal blocks and tapered slots are symmetrically arranged on the horn's top and lower constructions, as shown in Figure 4.28a. The top view of the bottom construction of the suggested horn antenna is depicted in Figure 4.28b. In particular, blocks 3, 4, and 5 are all the same size and are inserted next to the tapered slot to make the circular polarization (CP) function easier. In contrast, blocks 1 and 2 are designed for impedance matching and have different heights at the horn center. Additionally, to improve the impedance matching between the feeding waveguide and the flare area of the horn antenna, a stepped transition structure (STS) made up of blocks 1 and 2 is introduced at the horn's throat [15]. Table 4.10 provides specifics on the optimal dimensions.

4.4.3 Antenna Design and Analysis

4.4.3.1 Generation of the CP Operation (Ant. 1 to Ant. 2)

Based on an ideal linear polarization (LP) horn antenna, Ant. 1, which was built in compliance with the optimal horn design criteria, the design approach for the

Table 4.10 Dimensions of the Proposed Antenna.

Parameters	Value (mm)	Parameters	Value (mm)	Parameters	Value (mm)
a	15	w	60	g_2	10.3
m_1	2.9	h_p	4.765	g_3	5.4
m_2	2.6	h	9.53	g_4	4.4
p	5.4	h_1	0.3	g_5	6
d	22.5	h_2	1.4	l'	45
l_1	50	h_3	2.9	d'	30
l_2	30	g_1	4.9		

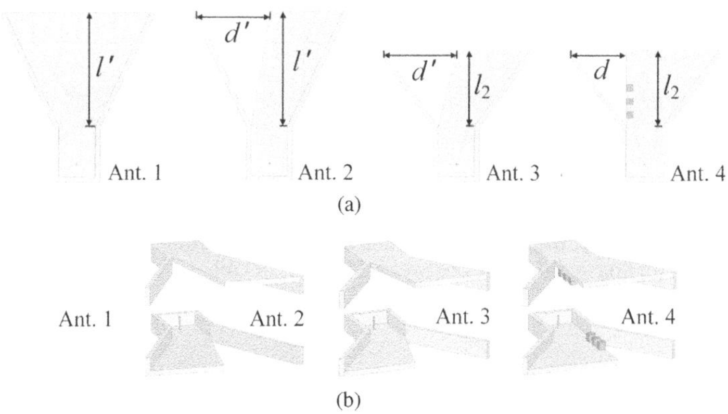

Figure 4.29 Evolution process of the CP horn antenna (Ant. 1: optimum LP horn; Ant. 2: CP horn with ATSs; Ant. 3: shortened horn without metal blocks; and Ant. 4: shortened CP horn with metal blocks). (a) Top view and (b) 3-D view. ($l' = 45$ mm, $d' = 30$ mm, $l_2 = 30$ mm, and $d = 22.5$ mm).

miniature CP horn antenna is presented in Figure 4.29. With the electric field component E_x vertical to the horn's broad wall, Ant. 1 illustrates LP operation. Ant. 2 is generated by cutting two ATSs of longitudinal lengths l' and d' on the horn's wide walls in order to create the orthogonal electric field component E_y, which is oriented parallel to the broad wall (Figure 4.30). Figure 4.31a,b compare the electric field distributions on Ant. 1 and Ant. 2's radiating apertures at 14.5 GHz. Cutting two ATSs yields the orthogonal electric field component E_y, which concentrates within the ATSs at $t = 0$ and $t = T/2$. Both E_y and E_x of Ant. 2 display practically comparable amplitude and quadrature phases, as demonstrated in Figure 4.30a,b. Ant. 2's 3-dB AR bandwidth of 30.36% in the frequency range of 11.23–15.25 GHz is validated by Figure 4.30c as a result of the applied ATSs. Nevertheless, Ant. 2's longitudinal dimensions are still large in spite of these developments.

4.4.3.2 Longitudinal Miniaturization of the Horn Antenna (Ant. 2 to Ant. 3)

The radiating aperture's size should be consistent for the best gain performance since the horn antenna's gain depends on it. Ant. 3's longitudinal dimension in Figure 4.30 is lowered from $l' = 45\,\text{mm}$ to $l_2 = 30\,\text{mm}$, a 33.3% reduction. Figure 4.31c shows the electric field distributions on the radiating aperture of Ant. 3 at 14.5 GHz. It shows that the amplitude of E_x increases at $t = T/4$ and $t = 3T/4$, while the amplitude of E_y decreases at $t = 0$ and $t = \text{T}/2$. As shown in Figure 4.30a, the amplitudes of E_y and E_x of Ant. 3 become uneven, which results in a degraded AR, as shown in Figure 4.30c.

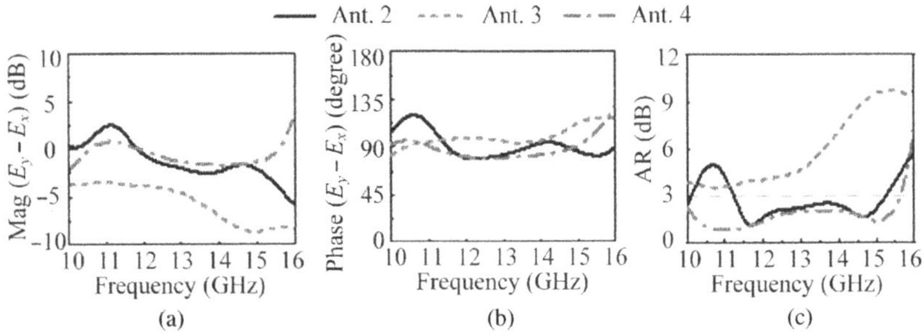

Figure 4.30 Comparisons of the Ant. 2, Ant. 3, and Ant. 4. (a) Amplitude differences between E_y and E_x. (b) Phase differences between E_y and E_x. (c) AR.

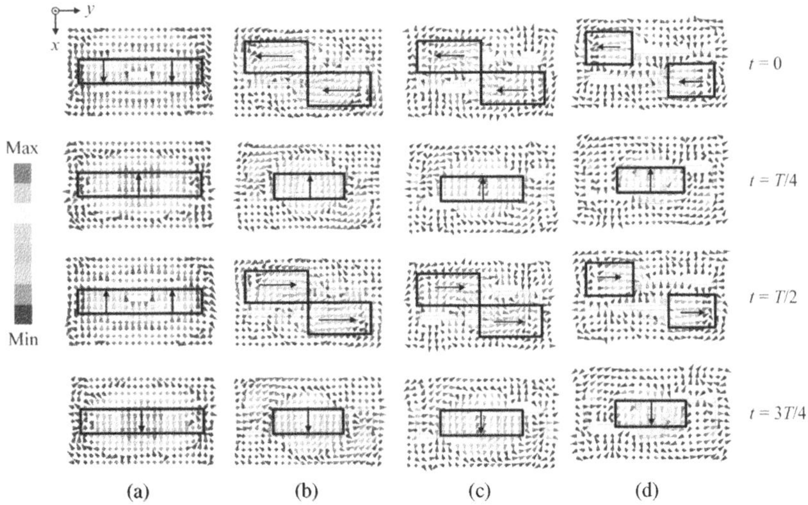

Figure 4.31 The comparison of electric field distributions on the radiating aperture at 14.5 GHz. (a) Ant. 1, (b) Ant. 2, (c) Ant. 3, and (d) Ant. 4.

4.4.3.3 CP Reconstruction in the Shortened Horn (Ant. 3 to Ant. 4)

Metal blocks that introduce additional capacitances are loaded close to the ATSs to obtain Ant. 4. This increases the capacitance between the two sides of the tapered slots, thereby adjusting the amplitude difference between the orthogonal electric field components $E\_(y)$ and $E\_(x)$ and reconstructing the CP operation in the shortened horn. For ATSs with straight edges, the lateral dimension is changed from d' to d. This makes it simpler to load the metal blocks and construct them using a conventional mechanical manufacturing technique.

The amplitude distributions of the electric fields adjacent to the tapered slots at 14.5 GHz for Ants 3 and 4 are presented in Figure 4.32. Stronger electric fields are focused near the slots owing to the implanted metal pieces. Consequently, the radiating apertures $E\_(y)$ amplitude rises (see $t = 0$ and $t = T/2$ in Figure 4.31d), and $E\_(x)$ amplitude lowers (see $t = T/4$ and $t = 3T/4$ in Figure 4.31d). While the phase difference between Ant. 4's $E\_(y)$ and $E\_(x)$ is around 90° over the working range, Figure 4.33b indicates that loading metal blocks provides virtually comparable

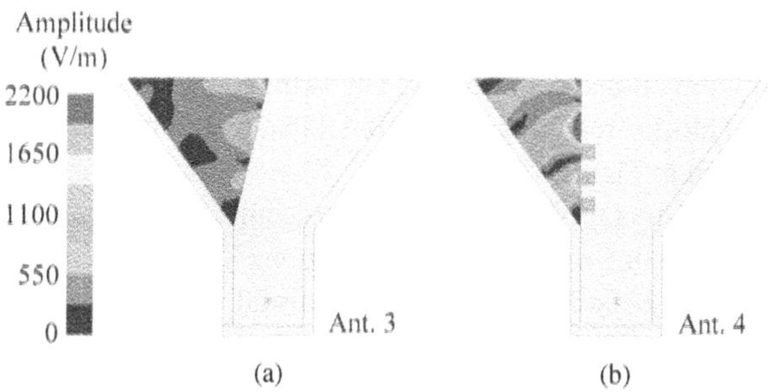

Figure 4.32 Amplitude distributions of electric field near tapered slots at 14.5 GHz. (a) Ant. 3 and (b) Ant. 4.

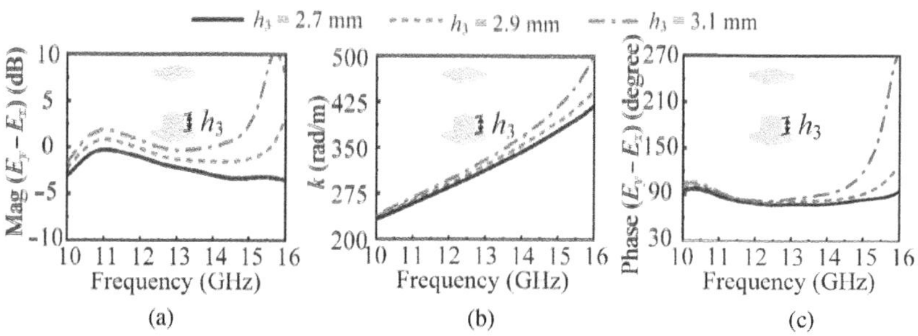

Figure 4.33 (a) Amplitude differences, (b) dispersion diagrams, and (c) phase differences for different metal block heights h_3.

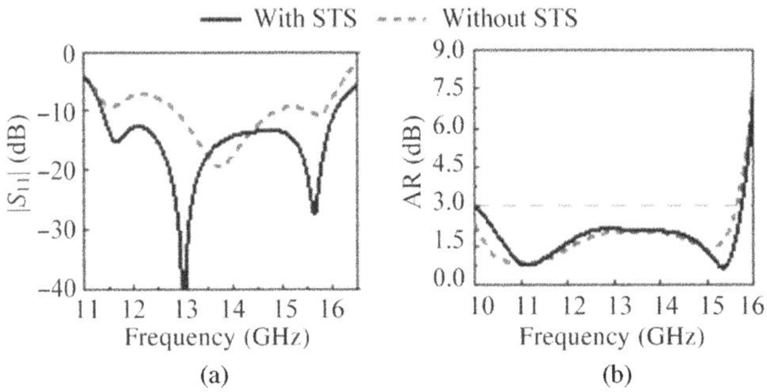

Figure 4.34 (a) $|S_{11}|$s and (b) AR plots of the antenna with or without STS.

amplitudes of $E\_(y)$ and $E\_(x)$. The CP equations stated below are met.

$$|E_x| = |E_y| \tag{4.13}$$

$$\phi_y - \phi_x = \pi/2 \tag{4.14}$$

where ϕ_y and ϕ_x are the phases of E_x and E_y, and $|E_x|$ and $|E_y|$ are the electric field's magnitudes in the x and y directions, respectively. The orthogonal polarizations are rebalanced using metal blocks, resulting in a 3-dB AR bandwidth of 44% from 10 to 15.64 GHz, as illustrated in Figure 4.33c.

Examining the height h_3 of metal blocks helps one to confirm how well the embedded blocks operate on the orthogonal polarizations. The capacitance between the metal block's side surface and the horn's thin wall will be enhanced by the rise in h_3. The capacitive effect causes the amplitude of E_y to expand as h_3 increases. Figure 4.34a demonstrates that as the height of the metal blocks h_3 expands, the amplitude difference increases as well.

Conversely, the inserted metal blocks increase the capacitance between the broad wall and the metal block's top surface. As a consequence, passing through the blocks affects the wave's propagation constant k [15]. As shown in Figure 4.35b, the increased parallel capacitive effect between the broad wall of the horn antenna and the top surface of the metal block causes the propagation constant to rise as the height of the metal block increases. The phase difference changes as the height of the metal pieces fluctuates, as shown in Figure 4.35c. When the metal block height is set suitably, the phase difference between E_x and E_y is around 90°, resulting in left-hand circular polarization (LHCP) radiation across a large range.

4.4.3.4 Improvement of Impedance Matching

To improve impedance matching between the feeding waveguide and the horn flare region, an STS made up of two pairs of metal blocks is integrated at the horn

Figure 4.35 (a) Upper structure, (b) lower structure, and (c) assembled antenna.

antenna's throat, as shown in Figure 4.29. Blocks 1 and 2 are intended to have optimal heights of 0.3 mm and 1.4 mm, respectively. To demonstrate how well STS works to enhance impedance matching, Figure 4.35a displays the magnitude of the S_{11} parameter for the CP horn antenna with and without STS. The impedance bandwidth increased from 14.37% (12.85–14.84 GHz) to 34.14% (11.37–16.05 GHz) with the aid of STS. Figure 4.35b indicates that the STS has a negligible influence on the CP operation. Analysis is done on how STS could improve impedance matching. Since the loaded STS contributes additional capacitance between the waveguide's upper and lower plates, the effective permittivity of the STS-laden region of the waveguide is raised using the following formulas [24]:

$$\varepsilon_{r\_eff} = \varepsilon_0 \cdot \frac{h}{2h_1} \tag{4.15}$$

Or

$$\varepsilon_{r\_eff} = \varepsilon_0 \cdot \frac{h}{2h_2} \tag{4.16}$$

ε_{r_eff} is the effective relative permittivity of the STS-loaded part. The wave impedance of the STS-loaded part of the waveguide can be calculated by the following equation:

$$Z_w = \frac{\sqrt{\mu_0/\varepsilon_{r\_eff}}}{\sqrt{1-(\lambda/2a)^2}} \tag{4.17}$$

Therefore, a stepped impedance transformer that improves impedance matching may be compared to the STS.

4.4.4 Simulation and Measurement Results

Aluminum alloy 6061 is used in the computerized numerical control (CNC) mechanical manufacturing method to create the suggested antenna. The images are shown in Figure 4.36. To excite the horn, a coaxial probe is placed into the rectangular waveguide's bottom structure.

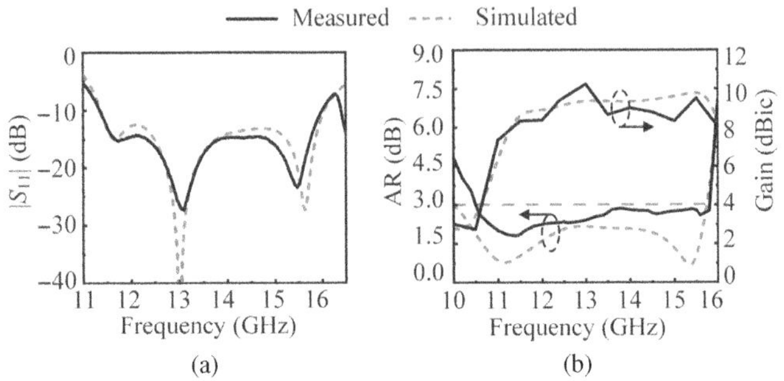

Figure 4.36 Simulated and measured results. (a) $|S_{11}|$ and (b) AR and gain.

Figure 4.36a indicates that the simulated and observed magnitudes of S_{11} match well. With $S_{11} < -10$ dB, the observed impedance bandwidth is 34% (11.35–16 GHz). The AR and gain plots of simulated and seen outcomes are displayed in Figure 4.37b. The 3-dB AR bandwidth measurement is 39.4% (10.6–15.8 GHz). The bandwidth of the recommended antenna is 32.8% (11.35–15.8 GHz) while taking into consideration both the impedance bandwidth and the 3-dB AR bandwidth for CP operation. 8.34–10.24 dBic is the range of the observed gain over the operating band. Figure 4.37 depicts the LHCP and right-hand circular polarization (RHCP) radiation patterns, which are constant across the working spectrum when simulated and observed.

According to Table 4.11's comparisons with other documented CP horns, the suggested antenna provides a greater CP bandwidth and a shorter longitudinal length than [25, 26]. Additionally, compared to the suggested CP horn antenna, the longitudinal lengths of the CP horn antennas in [27, 28] are much greater. As a result, the suggested horn antenna can effectively accomplish wideband CP functioning with a much reduced longitudinal length and a more straightforward design, which makes it suitable for applications that are sensitive to cost and space constraints.

4.4.5 Conclusion

In this letter, we suggest using metal blocks and ATSs to create a miniature wideband CP horn antenna. This is the first time a CP horn antenna's longitudinal profile has been decreased using embedded metal blocks. Consequently, our CP horn antenna is 33.3% smaller than the original CP horn, with a longitudinal dimension of just $1.36\lambda_0$ at the center frequency of the operational band. Furthermore, we get an appealing bandwidth of 32.8% (11.35–15.8 GHz) with a reflection coefficient $|S_{11}| < -10$ dB and an AR < 3 dB. Because of its cheap cost, wideband CP functioning, quick manufacture, and small and straightforward construction, this antenna design is ideal for automotive satellite communications (Figure 4.27).

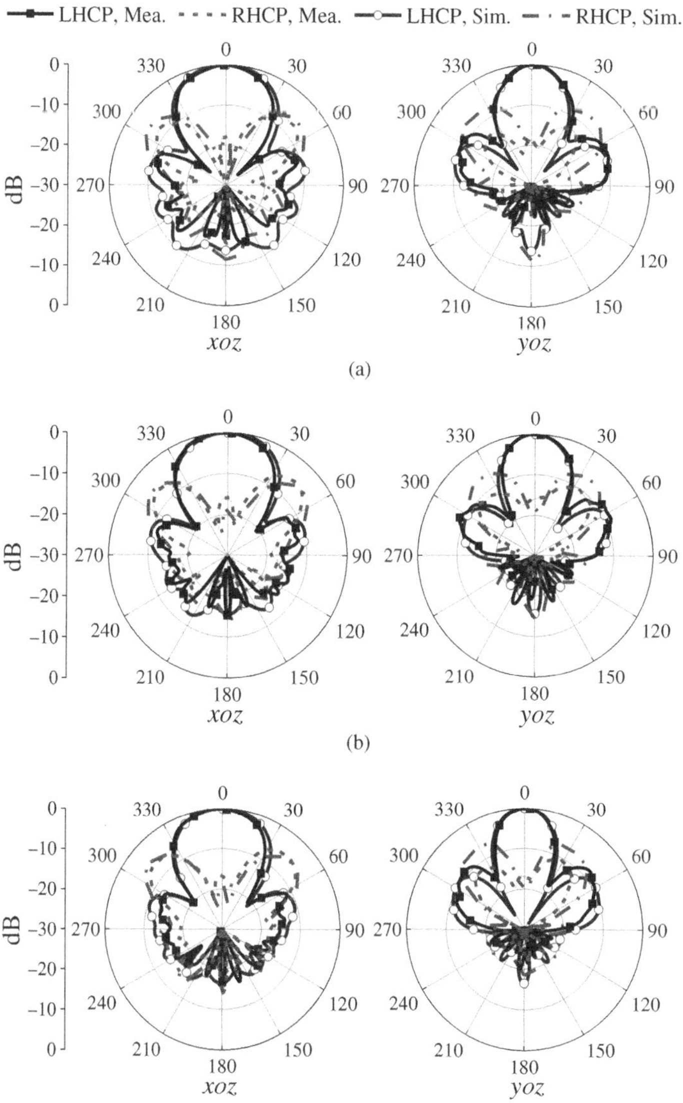

Figure 4.37 Simulated and measured normalized radiation patterns in the *xoz*-plane and the *yoz*-plane. (a) 12.5 GHz, (b) 13.5 GHz, and (c) 14.5 GHz.

Table 4.11 Comparison of Different CP Horn Antennas.

References	Frequency (GHz)	Length (λ_0)	Bandwidth[a)] (%)	Method of CP generation
[25]	29.6	3.01	7.4	Metasurface
[26]	10	2.83	26	SSPP
[27]	62.5	3.31	40 (AR < 3.2 dB)	ATSs and circular waveguide
[28]	112	9.64	37	Hexagonal waveguide
This work	13.5	1.36	32.8	ATSs and metal blocks

a) This is the overlap of S_{11} bandwidth and AR bandwidth.

4.5 Bidirectionally Fed Slow-Wave Substrate-Integrated Waveguide Monopulse Slot Array Antenna with Gain Enhancement

4.5.1 Introduction

An SW-SIW monopulse slot array antenna with gain improvement is described in our proposal. An SW-SIW leaky wave slot array, constructed of two layers of Rogers 5880 dielectric substrates, makes up the radiating part of the monopulse antenna. The SW-SIW, which has a shorter guided wavelength and a slowed phase velocity, is formed by raising the effective permittivity of the SIW by inserting periodic metalized via-holes in the lower substrate to give the requisite SW effect. With twice as many radiating unit cells as a characteristic SIW slot array with the same radiating aperture, the SW-SIW slot array delivers better radiation efficiency and gain. Using the same radiating aperture, we design a 20-unit SW-SIW monopulse slot array and compare it to a normal SIW monopulse slot array with only 10 units. The SWSIW monopulse slot array's sum and difference patterns exhibit gains of 2.5 and 2.1 dB, respectively, over the typical SIW monopulse slot array. We assure our design proposal by manufacturing and measuring a prototype of the SW-SIW monopulse slot array antenna.

The design and comparison of a typical SIW monopulse antenna with merely 10 units occupying the same radiating aperture with an SW-SIW monopulse slot array antenna with 20 radiating units are done. Sum and difference beams are produced by the recommended SW-SIW monopulse slot array antenna, which offers gains that are 2.5 and 2.1 dB greater than those of the standard SIW monopulse slot array antenna. With characteristics including high gain, affordability, and simplicity in production, our SW-SIW monopulse slot array antenna provides a competitive option for direction-finding applications.

4.5.2 Antenna Design

4.5.2.1 Slow-Wave Substrate-Integrated Waveguide

The SW-SIW is employed to produce the leaky wave slot array, which acts as the radiating component. The SW-SIW, which comprises two stacked substrates, is depicted in three dimensions in Figure 4.38. Table 4.12 provides the essential parameters. Rogers 5880, which has a dielectric loss tangent $\tan\delta$ of 0.0009 and a relative dielectric constant ε_r of 2.2, makes up the top and lower substrates. The symbols h_1 and h_2 represent the thicknesses of the top and bottom substrates, respectively. The SWS is achieved by including metalized blind via-holes with characteristics like diameter d_2, longitudinal period distance p_2, and lateral distance p_1 inside the lower substrate [6]. By increasing capacitance, this structure lowers the phase velocity v_p and raises the SIW's effective permittivity. The SW-SIW's guided wavelength can be changed in the manner described below:

$$\varepsilon_{r\_eff} = \varepsilon_r \left(\frac{h_1 + h_2}{h_1} \right) \tag{4.18}$$

$$v_p = \frac{1}{\sqrt{\mu_0 \varepsilon_{r\_eff}}} \tag{4.19}$$

$$\lambda_g = \frac{v_p}{f} \tag{4.20}$$

where ε_{r_eff} is the effective permittivity of SW-SIW.

By changing h_2, Figure 4.39 demonstrates how $\lambda_{g\_SIW}/\lambda_{g\_SW\text{-}SIW}$ fluctuates when $(h_1 + h_2)$ is constant. The findings reveal that when $(h_1 + h_2)$ remains constant, raising h_2 results in a decrease in the guided wavelength within the SW-SIW. Since they correspond to the substrates' commercially obtainable thicknesses, the values

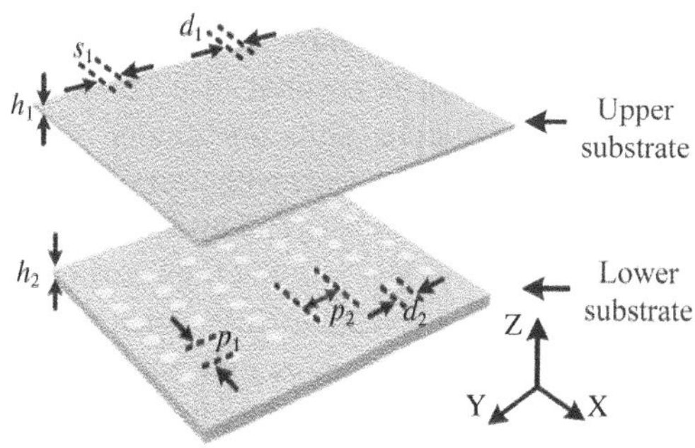

Figure 4.38 3-D exploded view of the SW-SIW.

Table 4.12 Dimensions of the SW-SIW (Units: Millimeter).

Variable	h_1	h_2	d_1	d_2	s_1	p_1	p_2
Value	0.254	0.787	0.6	0.9	1	1.6	2.5

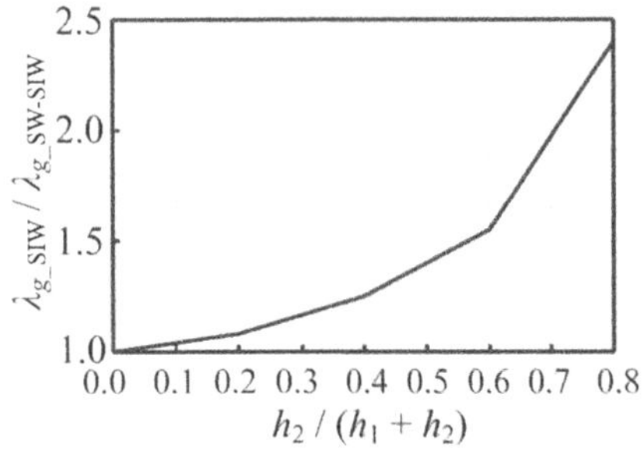

Figure 4.39 $\lambda_{g\_SIW}/\lambda_{g\_SW\text{-}SIW}$ versus $h_2/(h_1+h_2)$ at 12 GHz.

Table 4.13 Dimensions of the SW-SIW (Units: Millimeter).

Variable	d_3	g_1	g_2	w_1	w_2	w_3	w_4
Value	0.254	0.787	0.6	0.9	1	1.6	2.5
Variable	l_1	l_2	l_1	P_2	p_1	p_1	P_2
Value	0.254	0.787	0.6	0.9	1	1.6	2.5
Variable	p_1	p_2	l_1	w_2	d_1	d_1	p_2
Value	0.254	0.787	0.6	0.9	1	1.6	2.5

of h_1 and h_2 are eventually determined to be 0.254 mm and 0.787 mm, respectively. The $\lambda_{g\_SIW}/\lambda_{g\_SW\text{-}SIW}$ ratio is therefore roughly equal to 2, indicating half of the guided wavelength. In comparison to a traditional SIW slot array with an identical radiating aperture, the SW-SIW slot array can accommodate roughly twice as many radiating unit cells owing to the decrease in the guided wavelength, which increases the radiation efficiency and gain.

4.5.2.2 SW-SIW Leaky Wave Slot Array

In-phase radiation is ensured in a periodic leaky wave slot array by keeping neighboring radiating units one guided wavelength apart. In comparison to a conventional SIW slot array with a similar aperture, the SW-SIW's halved guided wavelength enables twice as many radiating unit cells. As shown in Figure 4.40a, a 20-unit SW-SIW slot array is thus created. By contrast, only 10 units may be accommodated by a standard SIW slot array (Figure 4.40b) that occupies the same aperture. Theoretically, more efficient energy radiation may be accomplished by increasing the number of units in a leaky wave array inside a given aperture. As shown in Figure 4.40c,d [29], a coaxial connection is developed to make the switch from the SW-SIW to a GCPW easier. Table 4.13 gives the SW-SIW slot array's precise measurements.

The electric field distributions of the leaky wave slot array based on SW-SIW and a conventional SIW at 12 GHz are compared in Figure 4.41 to clearly highlight

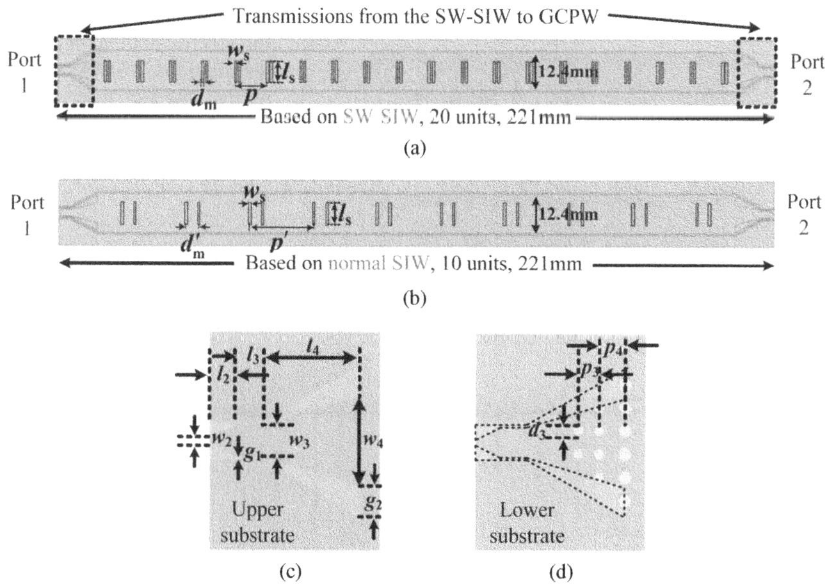

Figure 4.40 (a) Top view of the slot array based on the SW-SIW. (b) Top view of the slot array based on the normal SIW. (c) Top view of the transition for the SW-SIW slot array. (d) Bottom view of the transition for the SW-SIW slot array.

the SW effect. It is discovered that the guided wavelength in the SW-SIW is only half of the normal SIW. Two identical slots are used as radiating units to solve the broadside oblique slot boundary (OSB) issue [30, 31], producing two reflected waves with the same amplitude but opposite phase.

An SW-SIW and standard SIW slot arrays both operate in the 11–13-GHz region, as shown in Figure 4.42a. Leaky wave slot arrays made using SW-SIW have twice as many unit cells, which increases the power radiation across the slots and, eventually, the radiation efficiency. The amplification is reflected in the decreased transmission coefficient S_{21} of the leaky wave slot array, as shown in Figure 4.42a. The following equation [32] defines the overall efficiency η_{total} of the SW-SIW leaky wave slot array based on the S-parameters and the radiation efficiency η_{rad}:

$$\eta_{total} = \left(1 - |S_{11}|^2 - |S_{21}|^2\right) \eta_{rad} \tag{4.21}$$

can be improved as well, as Figure 4.42b illustrates.

By doubling the number of radiating unit cells in the array, SW-SIW considerably increases the directivity of the leaky wave slot array. When considering the leaky wave slot array's transmission coefficient (S_{21}) and reflection coefficient (S_{11}), which are determined by the following equation, this augmentation

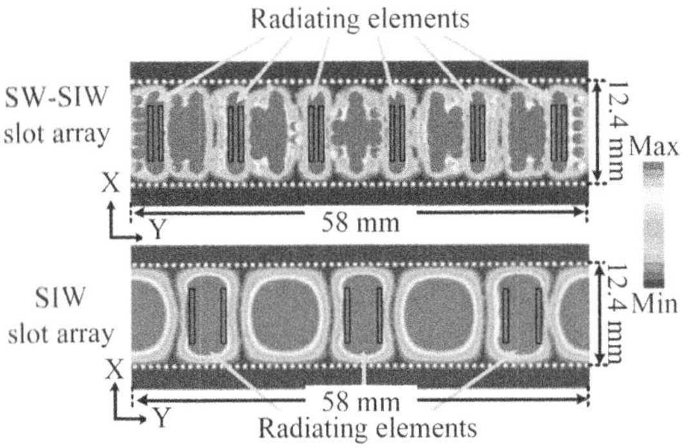

Figure 4.41 Comparison of the electric field magnitude distributions in SW-SIW and normal SIW slot arrays at 12 GHz.

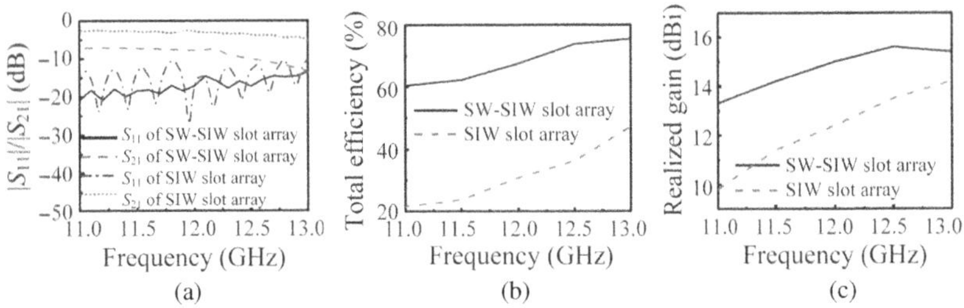

Figure 4.42 Comparisons of SW-SIW and normal SIW slot arrays. (a) *S*-parameters. (b) Efficiencies. (c) Gains.

is also seen in the actual gain:

$$G_{realized} = \eta_{total} \cdot D \tag{4.22}$$

can be improved as well; this is indicated by *d* (directivity). As shown in Figure 4.42c, the SW-SIW leaky wave slot array often has a higher realized gain than the conventional SIW slot array. In particular, using SW-SIW improves the realized gain by 2.6 dB at 12 GHz.

4.5.2.3 Principle of Bidirectionally Fed Sum/Difference Pattern

Figure 4.43 illustrates the bidirectionally fed leaky wave slot array idea with a sum/difference pattern. When port 1 excites the slot array and port 2 is filled with a 50-Ω load, the incident wave continuously radiates across the unit cells of the slot array, as shown in Figure 4.43a. According to the traveling wave antenna idea, the slot array would form a squint beam when the 50-Ω load absorbs the remaining incident traveling wave [17]. In contrast, as shown in Figure 4.43b, when port 1 is loaded with a 50-Ω load and port 2 excites the slot array, another squint beam is symmetrically created.

The two squint beams are superposed when ports 1 and 2 are excited simultaneously with opposing phases, as in Figure 4.43c, or with the same phase, as in

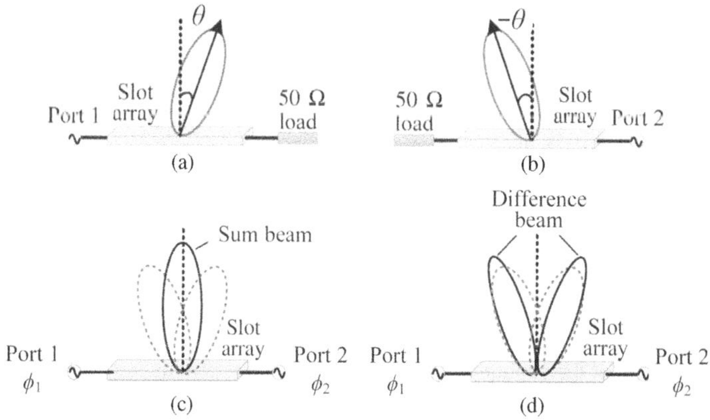

Figure 4.43 Principle of bidirectionally fed monopulse slot array. (a) Only port 1 is excited. (b) Only port 2 is excited. (c) Ports 1 and 2 are simultaneously excited with opposite phase, $\phi_1 = -\phi_2$. (d) Ports 1 and 2 are simultaneously excited with the same phase, $\phi_1 = \phi_2$.

Figure 4.43d, as shown by the following formula [33]:

$$F(\theta) = \left(F_a(-\theta) \cdot e^{j\phi_1} + F_a(-\theta) \cdot e^{j\phi_2}\right) \cdot F_u(\theta) \tag{4.23}$$

where $F(\theta)$ is pattern function of the bidirectionally-fed array antenna, $F_a(\theta)$ is the array function of the leaky wave slot array with single feed as shown in Figure 4.43(a), $F_a(-\theta)$ is the array function of the leaky wave slot array with single feed as shown in Figure 4.43(b), ϕ_1 and ϕ_2 present the phases of incident wave of ports 1 and 2, and $F_u(\theta)$ means the pattern of the unit slot. Formula (4.23) supposes the incident waves of ports 1 and 2 have same amplitude. So, the sum and difference beam can be realized by simultaneously exciting ports 1 and 2 with opposite phase ($\phi_1 = \phi_2$) or same phase ($\phi_1 = \phi_2$). The simulated sum and difference gain patterns of the bidirectionally-fed monopulse slot array based on SW-SIW and normal SIW are compared in Figure 4.44(a) and (b). By using SW-SIW, the gains of the sum and difference beams of the monopulse slot array are 2.5 and 2.1 dB higher than the normal SIW based array. The bandwidth of the sum/difference beam of the bidirectionally-fed monopulse SW-SIW slot array is 11.95–12.25 GHz, within which the sum and difference beams can be properly formed as shown in Figure 4.45.

To validate the design concept experimentally, a prototype of the proposed SW-SIW monopulse leaky wave slot array antenna is built and displayed in Figure 4.46a. The antenna's two ports are excited by a hybrid ring connected by two identical RF wires. The hybrid ring, the RF wires, and the leaky wave slot array are then each characterized independently using a vector network analyzer. The measured S-parameters of the leaky wave slot array without the hybrid ring

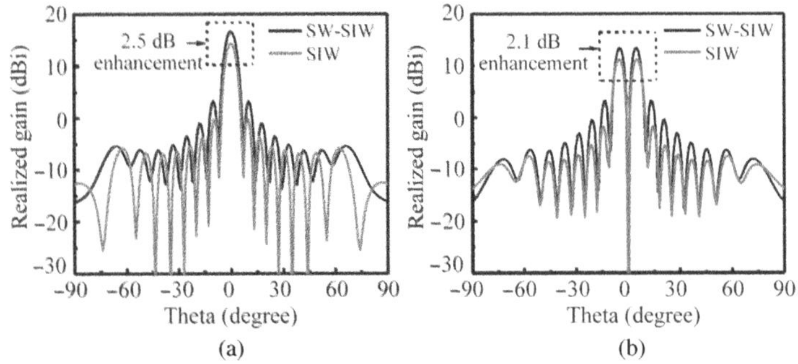

Figure 4.44 Comparison of the simulated gain patterns of bidirectionally fed monopulse slot array based on SW-SIW and normal SIW at 12 GHz. (a) Sum beam. (b) Difference beam.

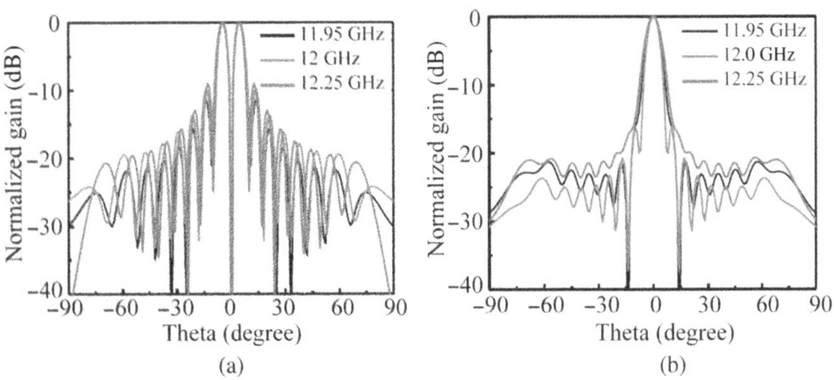

Figure 4.45 Normalized gain patterns of the bidirectionally fed monopulse SW-SIW slot array. (a) Sum beam. (b) Difference beam.

or RF wires connected are displayed in Figure 4.46b. Better than −11.6 dB and −10.7 dB, respectively, are the S_{11} and S_{21} magnitudes.

4.5.3 Result of Simulation and Measurement

To experimentally assess the design concept of the proposed SW-SIW monopulse leaky wave slot array antenna, a prototype is built and displayed in Figure 4.46a. The antenna's two terminals are excited by a hybrid ring connected to two identical RF cables. After that, a vector network analyzer is used to describe the RF wires, hybrid ring, and leaky wave slot array independently. Figure 4.46b displays the S-parameters of the leaky wave slot array without the hybrid ring and RF wires connected. The magnitudes of S_{11} and S_{21} are both better than −11.6 dB and −10.7 dB, respectively.

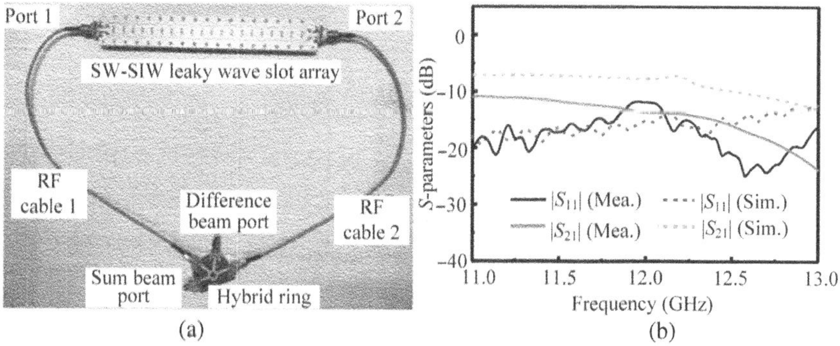

Figure 4.46 (a) Prototype of the proposed SW-SIW monopulse slot array antenna. (b) Measured *S*-parameters of the leaky wave slot array.

The *S*-parameters of the hybrid ring are displayed in Figure 4.47. As illustrated in Figure 4.47a, P_1 and P_2 represent the sum and difference beam ports, respectively. Furthermore, Figure 4.47b shows that the transmission phase errors for the difference and sum beams at 12 GHz are 5.2° and 2.7°, respectively. Additionally, in Figure 4.47c,d, the transmission coefficients S_{31}, S_{41}, S_{32}, and S_{42} are better than −3.5 dB, while P_1 and P_2's reflection coefficients, S_{11} and S_{22}, are better than −15 dB. Measurements of the two RF lines reveal insertion losses of less than 1.2 dB and nearly similar transmission phases.

A comparison of calculations and the actual sum and difference patterns of the proposed SW-SIW monopulse leaky wave slot array antenna is presented in Figure 4.48. The simulation's cumulative beam gain is 1.1 dB higher than the observed one, at 15.5 dBi. Similarly, the measured difference beam gain is 12.2 dBi, 1.4 dB lower than the simulation's prediction. The observed gain is calibrated by considering the insertion losses of the hybrid ring and the RF wires. The imbalance in the two peaks of the measured difference beam is caused by the hybrid ring's output amplitude and phase discrepancies. As illustrated in Figures 4.47b,d, the output phase and amplitude discrepancies of the hybrid ring's difference beam port P_2 are 4° and 0.22 dB, respectively.

Additionally, a comparison between the existing SIW monopulse antennas and the proposed SW-SIW monopulse slot array antenna is shown in Table 4.14. The proposed SW-SIW monopulse antenna has a relatively high aperture efficiency and the shortest half-power beamwidth (HPBW). The small beamwidth helps to improve the resolution of direction discoveries in radar tracking applications.

4.5.4 Conclusion

For a gain-enhanced monopulse antenna, this paper proposes an SW-SIW slot array with two layers of stacked dielectric substrates. Using the SW effect created

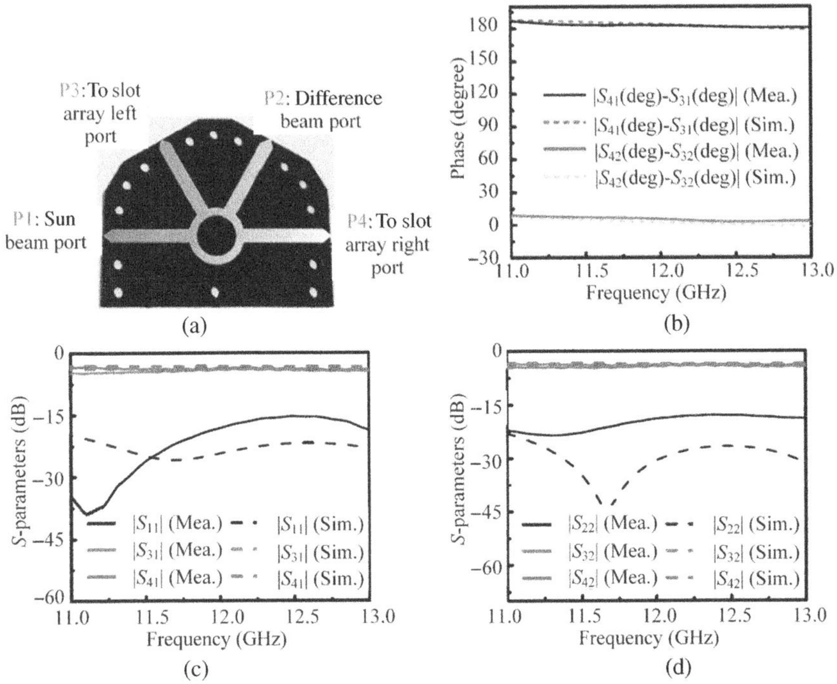

Figure 4.47 Hybrid ring and the S-parameters. (a) Photograph of the hybrid ring. (b) Transmission phase differences of P_1 and P_2. (c) Reflection and transmission coefficients of P_1. (d) Reflection and transmission coefficients of P_2.

Table 4.14 Dimensions of Horn Antennas.

References	Frequency (GHz)	Aperture size ($\lambda_0 \times \lambda_0$)	Gain (dBi)	Aperture efficiency (%)	HPBW
[34]	31.5	8.17×4.37	18.7	22	>15°
[35]	5.83	3.39×2.12	9.6	21	>15°
[36]	10	2.22×1.55	11.9	42	>30°
[37]	10	4.1×0.8	9.5	21	>30°
This work	12	8.84×0.8	15.5	39.9	8°

by periodic metalized via-holes placed in the bottom substrate, the guided wavelength is reduced by half. The SW-SIW slot array may have nearly twice as many radiating unit cells as a typical SIW slot array with the same radiating aperture in order to boost the radiation efficiency and gain. The experimental validation of a prototype SW-SIW monopulse slot array antenna demonstrates

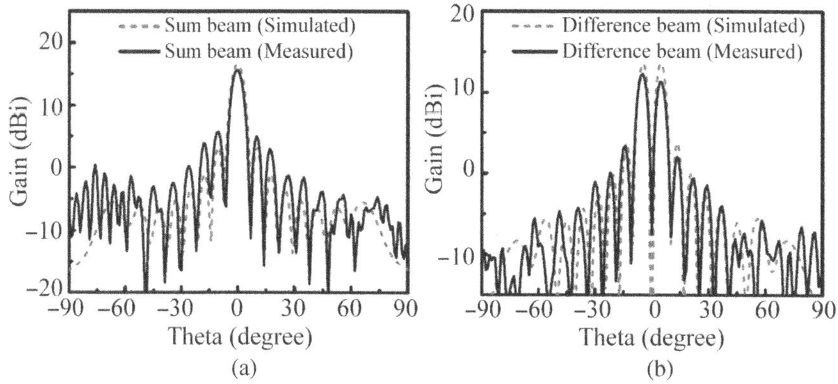

Figure 4.48 Gain patterns at 12 GHz. (a) Sum beams. (b) Difference beams.

the successful achievement of sum and difference patterns, offering a competitive solution for direction-finding applications with high gain, affordability, and ease of manufacture.

4.6 Compact Slow-Wave Half-Mode SIW Periodic Leaky Wave Antenna with Continuous Beam Scanning and High Gain

4.6.1 Introduction

This section proposes a compact periodic leaky wave antenna (PLWA) with continuous beam scanning and high gain based on an SW half-mode substrate-integrated waveguide (SW-HMSIW). The standard HMSIW is converted into the SW-HMSIW by including SWSs, which are composed of metalized blind vias and patches. This slows down the phase velocity of electromagnetic waves in the SIW, which leads to a 50% reduction in the guided wavelength. The SW-HMSIW PLWA cuts the spacing between radiating slots in half to allow twice as many radiating slots within the same length. This setup could greatly boost the PLWA's gain. Additionally, the patches of the SWS beneath the radiating slots are enlarged to enhance the impedance match of the radiating units and aid in suppressing the open-stopband (OSB). As a result, the radiation beam may continuously switch between forward and backward scanning. Across an operating frequency range of 13.7–15.7 GHz, a built and measured prototype of the proposed SW-HMSIW PLWA demonstrated a maximum gain of 11 dBi and a total scanning angle of 62.3°. The observed results show good agreement with the simulations.

In this work, we propose a PLWA with continuous beam scanning over the broadside based on the SW-HMSIW transmission line. Because the transverse slots serve as the radiating units and are separated by roughly a directed wavelength, the −1th space harmonic may radiate. SWSs, consisting of metallic blind vias and patches, effectively cut the directed wavelength in half. This makes it possible to have twice as many radiating slots in the same length as a normal HMSIW PLWA. The PLWA's gain consequently rises dramatically. Impedance matching is achieved by asymmetrically increasing the loaded patches of SWSs, effectively suppressing OSB. The use of HMSIW may also result in PLWAs having a smaller size. Experimental results indicate that the beam can continuously scan from backward to forward with a total scanning range of 62.3° and a maximum gain of 11 dBi.

4.6.2 Antenna Design

4.6.2.1 Configuration

The proposed SW-HMSIW PLWA's transverse cross section and 3-D schematic view are depicted in Figure 4.49a,b, respectively. The entire structure is built using two Rogers 4350B substrates with a relative dielectric constant.

$\tan\delta = 0.004$, a dielectric loss tangent, and $\varepsilon_r = 3.66$. The thickness of the top and bottom substrates is denoted by the letters h_1 and h_2, respectively. A bonding layer with a dielectric loss tangent ($\tan\delta = 0.004$) and a relative dielectric constant ($\varepsilon_r = 3.52$) exists between the top and bottom structures. The symbol for this layer's thickness is h_3. Metallic via-holes with a period of s_1 and a diameter of d_1 are embedded on both sides of the two substrates to form the HMSIW. Additionally, the bottom substrate contains two rows of metalized blind via-holes with a diameter of d_2 and a separation distance of s_2. Rectangular metal patches, l_p long and w_p wide, are printed on the top surface of the bottom substrate. The blind via-holes are then electrically connected to these patches. These loaded patches and metalized blind via-holes [38] comprise the SWS, which is distributed irregularly along the longitudinal direction with period s_2 to form the SW-HMSIW. Transverse radiating grooves measuring l_s long and w_s wide are present on the top surface of the top substrate. The width of the SW-HMSIW is denoted by the symbol w_1. The metal patches positioned beneath the radiating slots are extended by l_x.

The proposed PLWA is fed via a GCPW. As shown in Figure 4.49c, a transition from GCPW to SW-HMSIW is built using the process described in reference [29]. Table 4.15 lists the dimensions of the planned SW-HMSIW PLWA.

4.6.2.2 Gain Enhancement

As was previously said, the two main problems with different PLWAs are broadside OSB and poor gain with restricted aperture. To overcome the limiting aperture of the PLWA and improve its gain, the number of radiation units can be increased by lowering the period of the unit cell. The guided wavelength of the electromagnetic

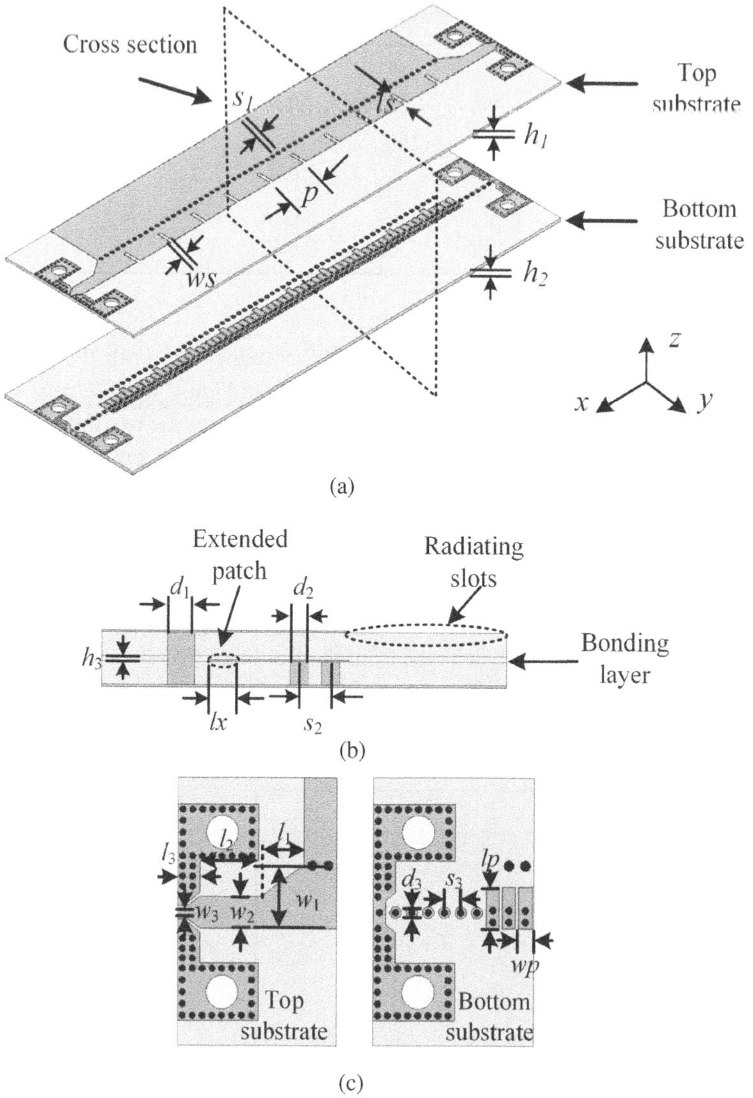

Figure 4.49 Geometry of the proposed SW-HMSIW PLWA. (a) Exploded 3-D view. (b) Transverse cross section. (c) Transition part.

wave traveling along SW-HMSIW is cut in half when using the SWS suggested in reference [38]. As a result, the radiation units are twice those of the HMSIW PLWA.

To evaluate the impact of the SWS, the traditional HMSIW PLWA and SW-HMSIW PLWA features are contrasted. The simulated gain of the conventional

Table 4.15 Parameters of Proposed PLWA (Units: Millimeters).

Variable	w_1	w_2	w_3	h_1	h_2	h_3	l_1
Value	3.8	1.9	0.5	0.508	0.508	0.12	2.7
Variable	d_1	d_2	d_3	s_1	s_2	s_3	l_2
Value	2.6	0.6	0.4	1	0.7	1	3.8
Variable	l_s	w_s	l_p	w_p	l_x	p	l_3
Value	2.4	0.5	2.5	0.8	0.59	6	1.4

HMSIW PLWA and the SWS-loaded HMSIW PLWA, which are both the same length but have twice as many radiating slots, is shown in Figure 4.50. Because of the OSB suppression in the SW-HMSIW PLWA, it is noted that the two antennas' operational frequency ranges differ. The next section presents a comprehensive examination of OSB suppression. A comparison of the two PLWAs is made possible by comparing their simulated antenna gain throughout the shared operational frequency range of 13.7–15.7 GHz. Because of the increased radiation energy brought about by the extra slots, Figure 4.50 demonstrates that the SW-HMSIW PLWA's realized gain is noticeably larger throughout the frequency range. However, since the energy leakage from the front and rear slots varies, the gain differential does not exactly show a twofold connection. However, larger gain is naturally caused by OSB suppression in the SW-HMSIW PLWA after around

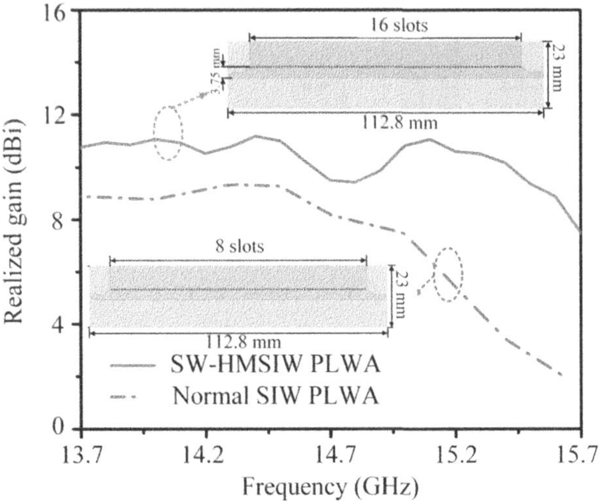

Figure 4.50 Comparison between simulated gain of normal HMSIW PLWA and SW-HMSIW PLWA.

15 GHz, when OSB manifests in the HMSIW PLWA. Consequently, adding more unit cells to PLWAs with a small aperture actually helps to increase gain.

4.6.2.3 OSB Suppression

It is possible to stimulate limitless space harmonics by carving periodic grooves along the SW-HMSIW. The leaking wave may be created by carefully designing the slots such that the −1th harmonic is radiated to free space. Theoretically, PLWAs' radiation beams may constantly scan in both forward and backward directions, including broadside. However, OSB, which results from the superposition of the energy reflected by unit cells, always happens in the broadside direction. Because of their often tiny magnitudes, the reflected power of the units may usually be ignored. However, OSB is present when all of the unit cells are in phase at the PLWA's input port, which causes the phase superimposition of the reflected waves from every unit cell.

The loaded patches underneath the slots are expanded to enhance the unit cells' impedance match and lessen reflected waves to minimize OSB. According to Figure 4.51a, the unit cell with a single transverse slit is an impedance-mismatched construction. When the loaded patches are extended, the unit cell's equivalent circuit, shown in Figure 4.51b, uses a series impedance Z in conjunction with a shunt admittance Y to achieve impedance matching.

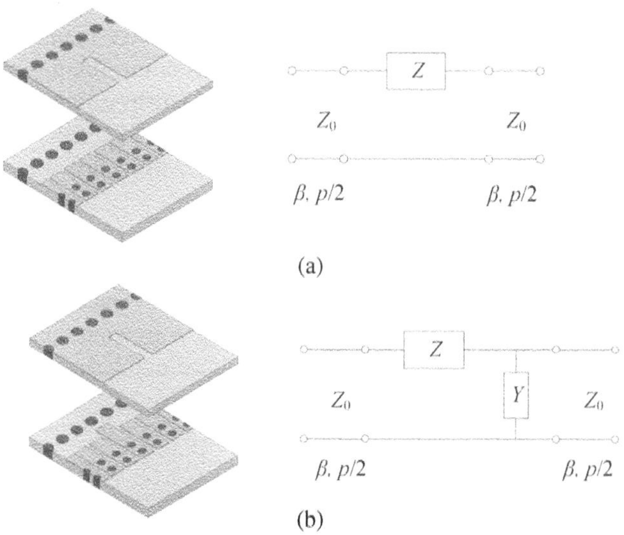

Figure 4.51 Equivalent circuits for different unit cells of SW-HMSIW. (a) Unit cell with only one slot. (b) Unit cell with both slot and extended patch.

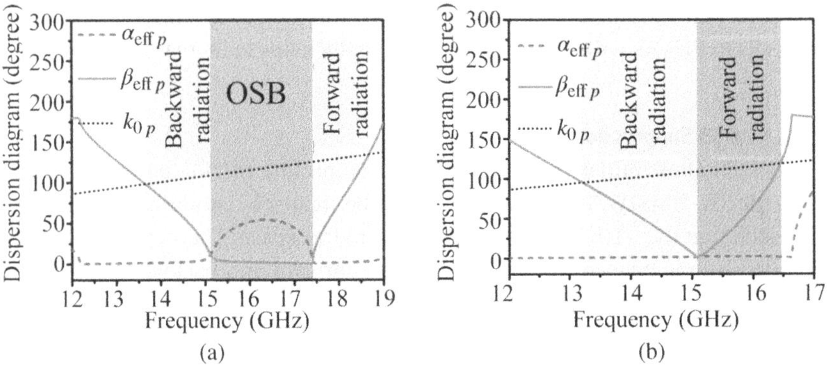

Figure 4.52 Simulated dispersion diagrams of different unit cells. (a) Initial SW-HMSIW PLWA unit cell shown in Figure 4.51a. (b) Proposed SW-HMSIW PLWA unit cell shown in Figure 4.51b.

The S-parameters of the equivalent circuit displayed in Figure 4.51b can be calculated as follows [39]:

$$S_{unit} = \begin{bmatrix} \dfrac{Z/Z_0 - YZ_0 + ZY}{2 + Z/Z_0 - YZ_0 + ZY} e^{-j\beta_{TL}p} & \dfrac{2}{2 + Z/Z_0 - YZ_0 + ZY} e^{-j\beta_{TL}p} \\ \dfrac{2}{2 + Z/Z_0 - YZ_0 + ZY} e^{-j\beta_{TL}p} & \dfrac{Z/Z_0 - YZ_0 + ZY}{2 + Z/Z_0 - YZ_0 + ZY} e^{-j\beta_{TL}p} \end{bmatrix} \tag{4.24}$$

Equation (4.24) states that under some conditions, the unit cell's S_{11} might drop to zero. As shown in Figure 4.51, the shunt admittance Y is directly determined by the length of the enlarged patch, l_x. According to Eq. (4.24), this shunt admittance Y influences the S-parameter. Therefore, S_{11} may be adjusted to zero by varying the patch length l_x. This means that in order to meet the condition that S_{11} becomes zero, the length of the expanded patch, l_x, must be set. OSB is less likely to develop when there is no reflected energy.

The aforementioned approach is validated by simulating dispersion characteristics to provide a more natural understanding. The unit cell's dispersion curves are found to be rather similar to the PLWA's with a lot of unit cells [40]. In Figure 4.52a, a huge OSB is shown, with β_{eff}, p being almost zero and α_{eff}, p growing significantly. It is clear from Figure 4.52b that OSB is indeed inhibited in the suggested SW-HMSIW unit cell. This confirms the fundamental design idea.

4.6.3 Conclusion

This study suggests a small PLWA with continuous beam scanning over the broadside that is based on an SW-HMSIW. To reduce the size of the antenna, the SWSs—which are made up of loaded patches and metalized blind via-holes—are added to HMSIW. The −1th space harmonic may radiate because transverse slots act as radiating units, spaced apart by about a directed wavelength. In order to enhance the impedance match of the radiating units and aid in OSB suppression, the extended loaded patches of the SWS underneath the radiating slots are also used. Because of this, the radiation beam may continually move from the rear to the front. Within the operational frequency range, the suggested SW-HMSIW PLWA attains a total scanning angle of 62.3° and a maximum gain of 11 dBi.

References

1 Wanzheng, L. (2004). Antenna theory and techniques. *Xi'an, China: Xi'an Electronic Science and Technology University*, 5, 153–154.

2 Mineo, M. and Paoloni, C. (2010). Corrugated rectangular waveguide tunable backward wave oscillator for terahertz applications. *IEEE Transactions on Electron Devices*, 57 (6), 1481–1484.

3 Mineo, M. and Paoloni, C. (2010). Double-corrugated rectangular waveguide slow-wave structure for terahertz vacuum devices. *IEEE Transactions on Electron Devices*, 57 (11), 3169–3175.

4 M. Mineo, C. Paoloni, A. Di Carlo, A. M. Fiorello, and M. Dispenza. (2009). Corrugated waveguide slow-wave structure for THz travelling wave tube. In *Proc. Eur. Microw. Conf. (EuMC)*, Sep. 2009, pp. 842–845.

5 Mineo, M. and Paoloni, C. (2010). Narrow corrugation rectangular waveguide for terahertz TWTs. *Electronics Letters*, 46 (13), 927–928.

6 Niembro-Martín, A. et al. (2014). Slow-wave substrate integrated waveguide. *IEEE Transactions on Microwave Theory and Techniques*, 62 (8), 1625–1633.

7 Ramaccia, D., Scattone, F., Bilotti, F., and Toscano, A. (2013). Broadband compact horn antennas by using EPS-ENZ metamaterial lens. *IEEE Transactions on Antennas and Propagation*, 61 (6), 2929–2937.

8 He, Y., Ding, N., Zhang, L., Zhang, W., and Du, B. (2015). Shortlength and high-aperture-efficiency horn antenna using low-loss bulk anisotropic metamaterial. *IEEE Antennas and Wireless Propagation Letters*, 14, 1642–1645.

9 Koerner, M.A. and Rogers, R.L. (2000). Gain enhancement of a pyramidal horn using E- and H-plane metal baffles. *IEEE Transactions on Antennas and Propagation*, 48 (4), 529–538.

10 Sun, D. and Xu, J. (2017). Compact phase corrected H-Plane horn antenna using slow-wave structures. *IEEE Antennas and Wireless Propagation Letters*, 16, 1032–1035.

11 Volakis, J.L. (2007). *Antenna Engineering Handbook*, 4th edn, McGrawHill, New York.

12 Coulombe, M., Nguyen, H.V., and Caloz, C. (2007). Substrate integrated artificial dielectric (SIAD) structure for miniaturized microstrip circuits. *IEEE Antennas and Wireless Propagation Letters*, 6, 575–579.

13 Wang, H., Fang, D.-G., Zhang, B., and Che, W.-Q. (2010). Dielectric loaded substrate integrated waveguide (SIW) H-plane horn antennas. *IEEE Transactions on Antennas and Propagation*, 58 (3), 640–647.

14 Wang, L., Esquius-Morote, M., Qi, H., Yin, X., and Mosig, J.R. (2017). Phase corrected H-plane horn antenna in gap SIW technology. *IEEE Transactions on Antennas and Propagation*, 65 (1), 347–353.

15 Deng, J.-Y., Luo, R.-Q., Lin, W. et al. (2021). Horn antenna with miniaturized size and increased gain by loading slow wave periodic metal blocks. *IEEE Transactions on Antennas and Propagation*, 69 (4), 2365–2369.

16 Esquius-Morote, M., Fuchs, B., Zürcher, J., and Mosig, J.R. (2013). A printed transition for matching improvement of SIW horn antennas. *IEEE Transactions on Antennas and Propagation*, 61 (4), 1923–1930.

17 Balanis, C.A. (2005). *Antenna Theory—Analysis and Design*, 3rd edn, Wiley, New York.

18 Deslandes, D. and Wu, K. (2006). Accurate modeling, wave mechanisms, and design considerations of a substrate integrated waveguide. *IEEE Transactions on Microwave Theory and Techniques*, 54 (6), 2516–2526.

19 Cai, Y., Qian, Z., Zhang, Y., Jin, J., and Cao, W. (2014). Bandwidth enhancement of SIW horn antenna loaded with air-via perforated dielectric slab. *IEEE Antennas and Wireless Propagation Letters*, 13, 571–574.

20 He, Y., Li, Y., Zhu, L., Bagci, H., Erricolo, D., and Chen, P. (2018). Waveguide dispersion tailoring by using embedded impedance surfaces. *Physical Review Applied*, 10 (6).

21 Mallahzadeh, A.R. and Esfandiarpour, S. (2018). Wideband H-Plane horn antenna based on ridge substrate integrated waveguide (RSIW). *IEEE Antennas and Wireless Propagation Letters*, 11, 85–88.

22 Liu, L. et al. (2020). Substrate integrated waveguide filtering horn antenna facilitated by embedded via-hole arrays. *IEEE Antennas and Wireless Propagation Letters*, 19 (7), 1187–1191.

23 Wang, L., Yin, X., Li, S., Zhao, H., Liu, L., and Zhang, M. (2014). Phase corrected substrate integrated waveguide H-Plane horn antenna with embedded metal-via arrays. *IEEE Transactions on Antennas and Propagation*, 62 (4), 1854–1861.

24 Zhang, Y., Deng, J.-Y., Sun, D., Yin, J.-Y., and Guo, L.-X. (2021). Compact slow-wave SIW H-plane horn antenna with increased gain for vehicular millimeter wave communication. *IEEE Transactions on Vehicular Technology*, 70 (7), 7289–7293.

25 Lin, C., Ge, Y., Bird, T.S., and Liu, K. (2018). Circularly polarized horns based on standard horns and a metasurface polarizer. *IEEE Antennas and Wireless Propagation Letters*, 17 (3), 480–484.

26 Li, D., Jiao, Y., Yu, H., and Weng, Z. (2021). Wideband circularly polarized pyramidal horn antenna based on spoof surface plasmon polaritons. *IEEE Transactions on Antennas and Propagation*, 69 (4), 2353–2358.

27 Cheng, X., Yao, Y., Yu, T., Chen, Z., Yu, J., and Chen, X. (2018). Analysis and design of a low-cost circularly polarized horn antenna. *IEEE Transactions on Antennas and Propagation*, 66 (12), 7363–7367.

28 Bhardwaj, S. and Volakis, J.L. (2018). Hexagonal waveguide based circularly polarized horn antennas for sub-mm-wave/terahertz band. *IEEE Transactions on Antennas and Propagation*, 66 (7), 3366–3374.

29 R. Kazemi, A. E. Fathy, S. Yang, and R. A. Sadeghzadeh. (2012). "Development of an ultra wide band GCPW to SIW transition". In *Proc. IEEE Radio Wireless Symp.,* 2012, pp. 171–174.

30 Lyu, Y.-L. et al. (2016). Leaky-wave antennas based on noncutoff substrate integrated waveguide supporting beam scanning from backward to forward. *IEEE Transactions on Antennas and Propagation*, 64 (6), 2155–2164.

31 Liu, J., Zhou, W., and Long, Y. (2018). A simple technique for open-stopband suppression in periodic leaky-wave antennas using two nonidentical elements per unit cell. *IEEE Transactions on Antennas and Propagation*, 66 (6), 2741–2751.

32 Liu, J. and Liang, J. (2021). Gain enhancement of transversely slotted substrate integrated waveguide leaky-wave antennas based on higher modes. *IEEE Transactions on Antennas and Propagation*, 69 (8), 4423–4438.

33 Topak, E., Hasch, J., Wagner, C., and Zwick, T. (2013). A novel millimeter-wave dual-fed phased array for beam steering. *IEEE Transactions on Microwave Theory and Techniques*, 61 (8), 3140–3147.

34 Liu, B. et al. (2009). Substrate integrated waveguide (SIW) monopulse slot antenna array. *IEEE Transactions on Antennas and Propagation*, 57 (1), 275–279.

35 Cao, F., Yang, D., Pan, J., Geng, D., and Xiao, H. (2017). A compact single-layer substrate-integrated waveguide (SIW) monopulse slot antenna array. *IEEE Antennas and Wireless Propagation Letters*, 16, 2755–2758.

36 Yang, T., Zhao, Z., Yang, D., Liu, X., and Liu, Q.-H. (2019). A single-layer SIW slots array monopulse antenna excited by a dual-mode resonator. *IEEE Access*, 7, 131282–131288.

37 Meiguni, J., AminKhatami, S., and Amneelahi, A. (2018) Compact substrate integrated waveguide mono-pulse antenna array. *International Journal of RF and MicrowaveComputer-Aided Engineering*, 28 (1), e21155.

38 Zhang, Y., Deng, J.Y., Sun, D.Q., Yin, J.Y., Guo, L.X., Ma, X.H., and Hao, Y. (2021). Slow wave substrate-integrated waveguide with miniaturized dimensions and broadened bandwidth. *IEEE Transactions on Microwave Theory and Techniques*, 69 (8), 3675–3683.

39 Cao, W., Chen, Z.N., Hong, W., Zhang, B., and Liu, A. (2014). A beam scanning leaky-wave slot antenna with enhanced scanning angle range and flat gain characteristic using composite phase-shifting transmission line. *IEEE Transactions on Antennas and Propagation*, 62 (11), 5871–5875.

40 Zhou, W., Liu, J., and Long, Y. (2018). Investigation of shorting vias for suppressing the open stopband in an SIW periodic leaky-wave structure. *IEEE Transactions on Microwave Theory and Techniques*, 66 (6), 2936–2945.

5

Antenna Array Based on SW-SIW

5.1 Miniaturized Slow-Wave Substrate-Integrated Waveguide Rotman Lens Compact Multibeam Antenna for Satellite-Assisted Internet of Vehicles

5.1.1 Introduction

For Internet of Vehicles (IoV) networks, low Earth orbit (LEO) satellite constellations, like SpaceX's Starlink, may provide reasonably priced worldwide satellite communication services. Even if users are far apart geographically, these satellites enable them to communicate data at the same time. Multibeam antennas are necessary for ground vehicles to link to many LEO satellites. For Internet of Things (IoT) cars to connect to accessible satellites, multibeam antenna arrays are essential, especially when the cars are on the ground and moving. According to Figure 5.1's application diagram, satellite-assisted IoV applications strongly favor small and inexpensive multibeam antennas. Since passive multibeam antennas do not need pricey transmit/receive (TR) modules made up of digital phase shifters, low-noise receivers, and power amplifiers, they are more affordable than active multibeam antennas. Because they provide reliable beam aiming, wide-angle coverage, and straightforward construction, Rotman lenses—also referred to as beam-forming networks—are used for passive multibeam antennas.

To provide a tiny and affordable multibeam antenna for satellite-assisted IoV applications, a miniaturized Rotman lens based on a slow-wave (SW) substrate-integrated waveguide (SIW) is described for the first time. This is achieved by embedding SW structures in the form of metalized blind holes inside the lens body, which lowers the guided wavelength of wave propagation within the lens. In contrast, the tiny Rotman lens body is 66.1% smaller than the conventional SIW Rotman lens. The quite satisfactory measured results of

Slow-Wave Substrate-Integrated Waveguide: Designing of Devices and Antennas, First Edition.
Jing-Ya Deng, Jia-Yuan Yin, Fengxia Li, Yulong Liu, Li-Xin Guo, and Xiao-Hua Ma.

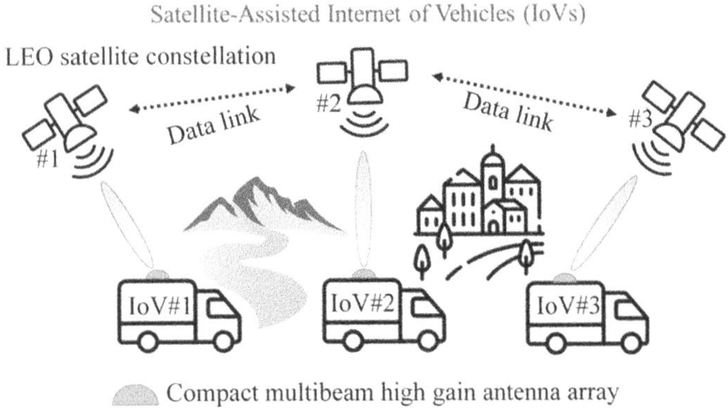

Figure 5.1 Multibeam antenna for LEO-satellite-assisted IoV applications.

a five-beam antenna based on the miniature Rotman lens confirms the design concept. The prototype was developed in the Ku-band, which is frequently utilized for satellite applications. The achieved performance confirms that the designed small SW-SIW Rotman lens is ideal for satellite-assisted IoV applications.

5.1.2 The Miniaturized Rotman Lens Design

5.1.2.1 Conventional SIW Rotman Lens

As shown in Figure 5.2, a conventional SIW Rotman lens working at 15.8 GHz is constructed using the Rogers 5880 dielectric substrate, which has a relative dielectric constant $\varepsilon_r = 2.2$ and a dielectric loss tangent $\tan\delta = 0.0009.2$. The focal angle α is 30°, and the focal length off-axis is F. The on-axis focal length G is 39.3 mm. The ratio of the on-axis to off-axis focal length can be found. g, as

$$g = \frac{G}{F} = 1.1 \tag{5.1}$$

To prevent the lens's performance from degrading, two fake ports, D_1 and D_2, are made to absorb unwanted electromagnetic radiation. In Figure 5.2, the greatest distance between two places on the lens body's contour (designated as W) is 46 mm. The larger lens body section, which takes up most of the antenna area and is surrounded by a yellow line in Figure 5.2, has an area of 1462.2 mm^2. For IoV applications that need a small platform, this architecture is not appropriate. To prevent duplication, the design principles and processes for a traditional SIW Rotman lens were covered in-depth in Reference [1] and will not be revisited here.

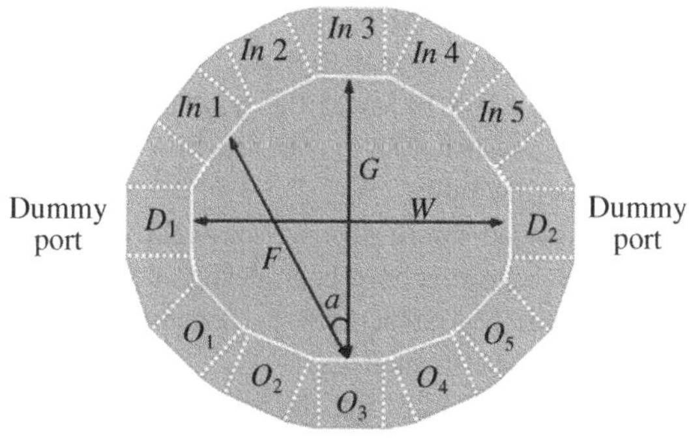

Figure 5.2 Standard SIW Rotman lens configuration. (The yellow line surrounds the lens body, which has a size of 4.1 λ_0^2).

5.1.2.2 Traditional Rotman Lens Miniaturization Methods

In previous studies, two conventional methods were employed to produce small Rotman lenses. The first approach is to design shorter input or output transmission lines inside the Rotman lens. For example, in References [1, 2], the longer equal-width delay lines were replaced with shorter unequal-width phase shifters. In order to achieve downsizing, Chebyshev impedance transformers were used for all ports in Reference [3]. Compared to the conventional linearly tapered transformers, these transformers were shorter. Furthermore, in Reference [4], the lengthy triangular transitions of the typical Rotman lens with the energy distribution slots were truncated, resulting in shorter input and output transmission line lengths. Power dividers were included with the beam ports to create a Rotman lens in Reference [5]. Nevertheless, the lens bodies of References [1–5] remained huge and were not reduced in size.

The traditional Rotman lenses are folded or wrapped as in the second technique. This design idea was used in Reference [6] to wrap Rotman lenses with a unique conformal substrate or fold them along the central plane. Despite the effective 50% miniaturization of the overall area of these folded Rotman lenses, implementation and fabrication are very challenging, which may be why the designs in References [6, 7] did not include any fabrication or measurable findings. Additionally, in References [8, 9], multibeam antennas were made smaller by stacking the radiating components, phase shift structures, and Rotman lens at various dielectric layers. Nonetheless, the Rotman lens itself, which occupies the majority of the multibeam antennas' surface, remained enormous because the waves inside it travel over a wide range of wavelengths. This is due to the fact that the Rotman lens needs to take up an equivalent amount of room. This finding led to the hypothesis that if the directed wavelength within the Rotman lens was much decreased, it might be made smaller. In this context, the SW-SIW offers a viable remedy.

5.1.2.3 Slow-Wave SIW

In order to create smaller SIWs, surface wave structures have recently been included in SIWs to lower the guided wavelength and phase velocity [10–12]. This is accomplished by raising the SIW's equivalent relative permittivity and/or relative permeability.

Miniaturized microwave devices [13] and antennas [14, 15] with drastically smaller dimensions have been designed using the SW-SIW idea. A Rotman lens incorporates SW structures to accomplish reduction without compromising transmission performance.

As shown in Figure 5.3, the SW-SIW is composed of two stacked Rogers 5880 dielectric substrates. Through-holes on the bottom substrate that are periodically metalized produce an SW structure that increases the effective relative permittivity of the SIW. By choosing suitable values for the longitudinal period distance p_2, the lateral distance p_1 between the through-holes, and the ratio h_2/h_1 of the substrate thicknesses, the SW effect of the SW-SIW may be customized [10]. When building a miniature Rotman lens using SW-SIW, these factors are regarded as degrees of freedom. With the overall thickness $h_2 + h_1$ being constant at 1.041 mm, decreasing the values of p_1, p_2, and h_2 results in a greater capacitance between the top surface of the upper substrate and the top plate of the SW structure. As a consequence, SW-SIW's capacitance per unit length increases, raising the effective relative permittivity ε_{r_eff}. Figures 5.4(a–c) show these phenomena, respectively. As a consequence, the guided wavelength may be greatly shortened and the phase velocity can be greatly decreased, as shown by the following equations:

$$\varepsilon_{r\_eff} = \varepsilon_r \cdot \left(\frac{h_1 + h_2}{h_1}\right) \tag{5.2}$$

$$v_p = \frac{c_0}{\sqrt{\mu_r \cdot \varepsilon_r \cdot \left(1 - \left(\frac{f_c}{f}\right)^2\right)}} \tag{5.3}$$

$$\lambda_g = \frac{v_p}{f} \tag{5.4}$$

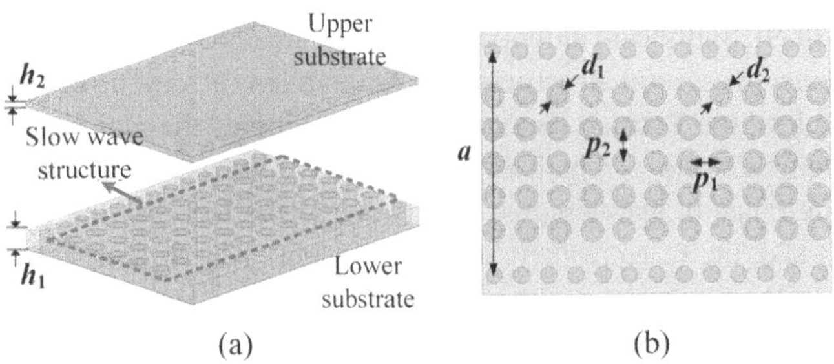

Figure 5.3 Geometry of SW-SIW. (a) Exploded 3D view. (b) Top view of the lower substrate ($a = 8$ mm, $d_1 = 0.6$ mm, $d_2 = 0.84$ mm, and $h_1 + h_2 = 1.041$ mm).

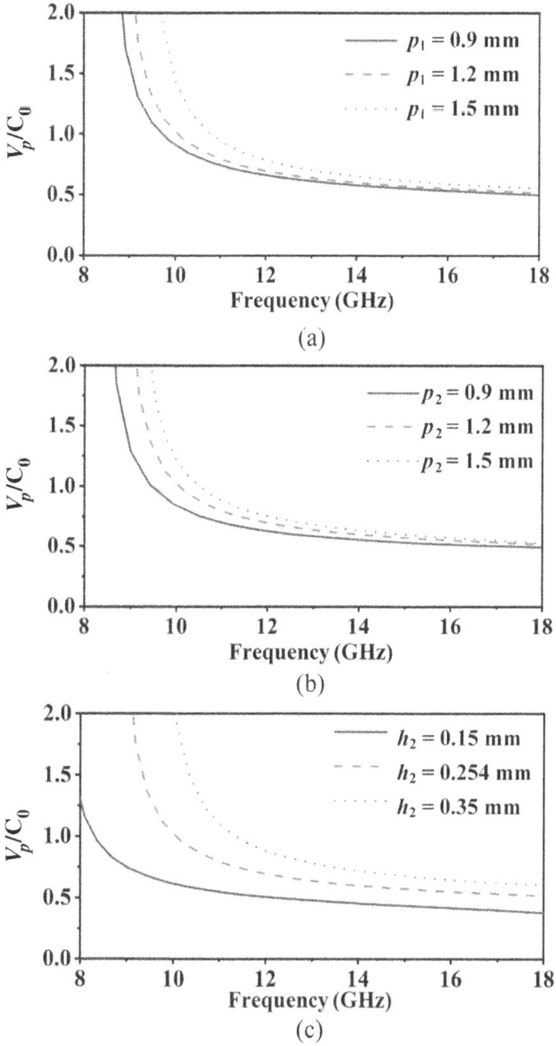

Figure 5.4 Normalized phase velocity for (a) p_1, (b) p_2, and (c) h_2 as a function of frequency.

where $\varepsilon_{r_{eff}}$ is the effective permittivity of the SW-SIW, $v(p)$ is the phase velocity, c_0 is the speed of light in free space, f_c is the cut-off frequency of the SW-SIW, and λ_g is the guided wavelength in SW-SIW.

5.1.2.4 Miniaturized Rotman Lens Body Based on SW-SIW

By adding SIW structures to SW-SIW, the guided wavelength can be greatly decreased. This capability can be used to miniaturize the lens body by incorporating a capacitance-enhanced SW structure in the form of periodic metalized via holes in the bottom substrate. This is necessary for multibeam-forming functionality and is shown by a yellow line in Figure 5.2. Figure 5.5 illustrates

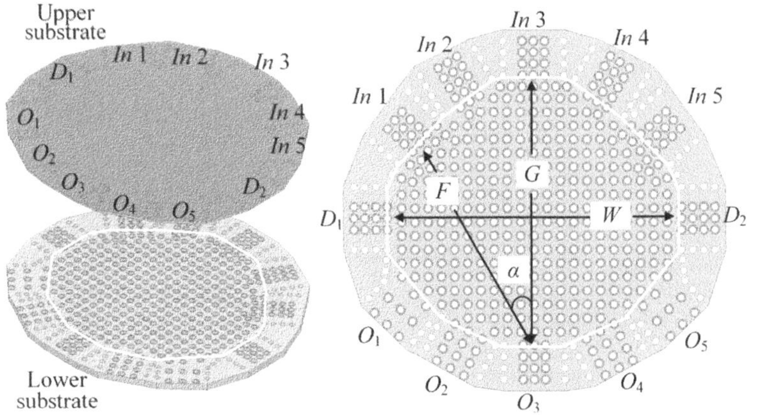

(a) (b)

Figure 5.5 Suggested miniature Rotman lens configuration. (The region that the yellow line surrounds is the lens body.) (a) Exploded 3D perspective. (b) Lower substrate from above.

the construction of a tiny Rotman lens using two layers of Rogers 5880 dielectric substrate. The thickness of the higher substrate (h_2) is 0.254 mm, whereas that of the bottom substrate (h_1) is 0.787 mm. Focus angle α is 30°, off-axis focal length F is 20.3 mm, and on-axis focal length G is 23 mm. The farthest distance between two locations on the contour of the lens body (shown as W) in Figure 5.5 is 26.6 mm, and the diameter of all the via holes is 0.6 mm. The longitudinal and transverse distances between the lens's SW metalized blind holes are 1.2 mm. The lens body area (delineated by the yellow lines in Figures 5.5a, b) is approximately 495.4 mm^2, which is 66.1% smaller than the lens body of the typical SIW Rotman lens in Figure 3.2. To construct the small SW-SIW Rotman lens, the electrical lengths of the wave channels for each port inside the lens body are taken to be equal to those in the conventional SIW Rotman lens. In theory, the smaller design's output phases ought to be almost the same as the larger design's ones.

The electric field magnitude distributions of the small Rotman lens and the standard Rotman lens are contrasted in Figure 5.6. Generally speaking, the field distributions and wave path conditions in the tiny SW-SIW Rotman lens body are similar to those in the conventional SIW Rotman lens. However, lens bodies G and W have a 42.2% (from 46 to 26.6 mm) and 41.5% (from 39.3 to 23 mm) reduction in actual form, respectively. The area of the lens body shrinks by 66.1%, from 1462.2 to 495.4 mm^2. Despite having smaller physical dimensions, the electric field distributions of the miniaturized SW-SIW-based Rotman lens are comparable to those of the conventional SIW Rotman lens, indicating how effectively the SW structure allows for miniaturization without compromising the desired field distributions.

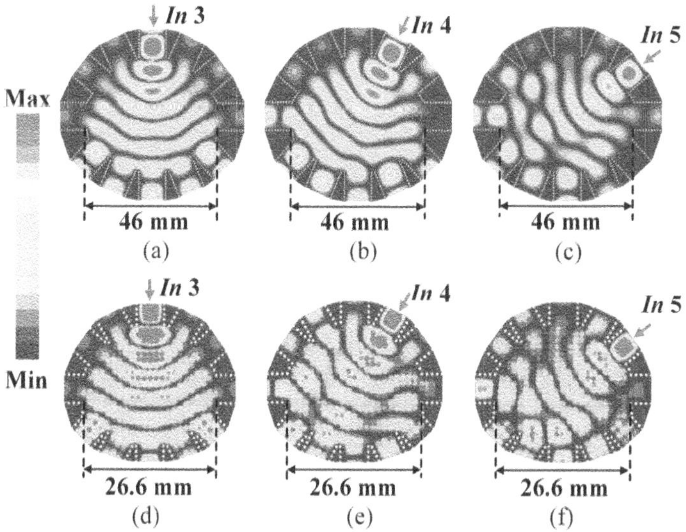

Figure 5.6 *E*-field magnitude distributions of the miniaturized Rotman lens and conventional Rotman lens. (a) Conventional Rotman lens fed from *In* 3. (b) Conventional Rotman lens fed from *In* 4. (c) Conventional Rotman lens fed from *In* 5. (d) SW-SIW Rotman lens fed from *In* 3. (e) SW-SIW Rotman lens fed from *In* 4. (f) SW-SIW Rotman lens fed from *In* 5.

5.1.2.5 Design of the Phase Shifters

For multibeam antenna applications, maintaining stable multibeam directions requires precise phase outputs from the Rotman lens. As shown in Figure 5.7, the outputs of the Rotman lens (O_1–O_5) must be connected to the inputs of the antennas (*Out* 1–*Out* 5) using transmission lines. These SW-SIW-based transmission lines must offer steady phase transitions, despite their varying actual lengths. Because of this, they function as phase shifters. It is possible to slow down the wave's phase velocity in SW-SIW and adjust the phase shifters by changing the transverse and longitudinal spacing of the metalized blind holes, as shown in Figure 5.4. By altering the transverse and longitudinal spacings of the metalized blind holes in the SW-SIW, the phase shifts for the different antenna inputs (*Out* 1–*Out* 5) can be changed, as shown in Figure 5.7c.

Since the micro-Rotman lens is symmetrical with respect to *Out* 3, Figure 5.7c only displays the phase shifters connected to *Out* 1, *Out* 2, and *Out* 3. As the physical lengths of the SW-SIW connected to O_1, O_2, and O_3 gradually decrease, the SW factors must be increased proportionately to ensure a steady output phase. The periods of the metalized blind holes in the SW transmission lines connected to *Out* 1, *Out* 2, and *Out* 3 have been optimized, as shown in Figure 5.7c.

Figure 5.8 shows the normalized phase velocities as functions of frequency. The SW-SIW phase shifter that is linked to *Out* 1 has the quickest phase velocity, while

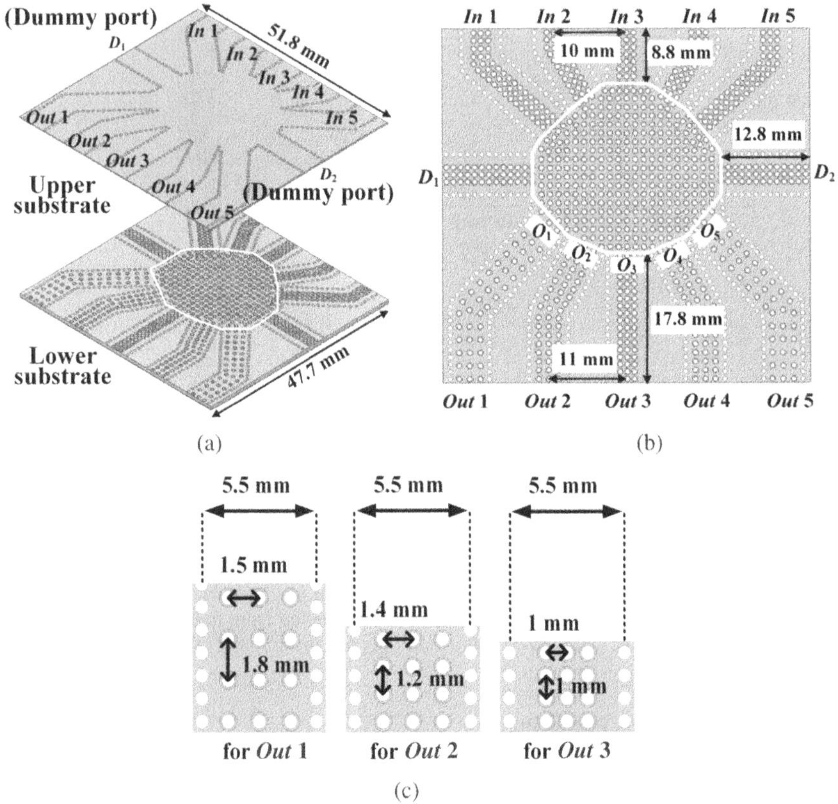

Figure 5.7 Configuration of the proposed miniaturized Rotman lens. (a) Exploded 3D view. (b) Top view of the lower substrate. (c) Optimized SW structures with different SW factors.

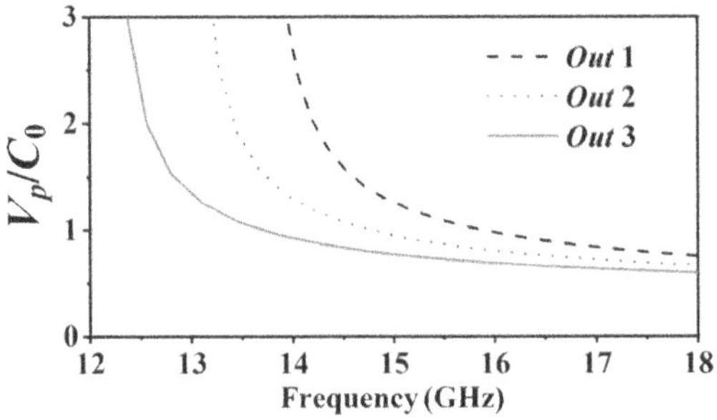

Figure 5.8 Normalized phase velocity of SW-SIW phase shifters connected with *Out* 1–*Out* 3.

the one that is connected to *Out* 3 has the slowest phase velocity. This guarantees that even when the phase shifters' physical lengths vary, they produce the same phase change.

5.1.2.6 Miniaturized SW-SIW Rotman Lens

The shrunken SW-SIW Rotman lens is accomplished by effectively cascading the SIW-based phase shifters, the small SW-SIW-based Rotman lens body, and the SW-SIW feeding lines, as shown in Figures 5.7a, b. This configuration ensures that different components are correctly connected and positioned in the Rotman lens to function as intended. To evaluate the performance of the full micro-Rotman lens connected to SW-SIW feeding lines and phase shifters, Figure 5.9 displays the electric field distribution of the entire lens fed by ports *In* 3, *In* 4, and *In* 5.

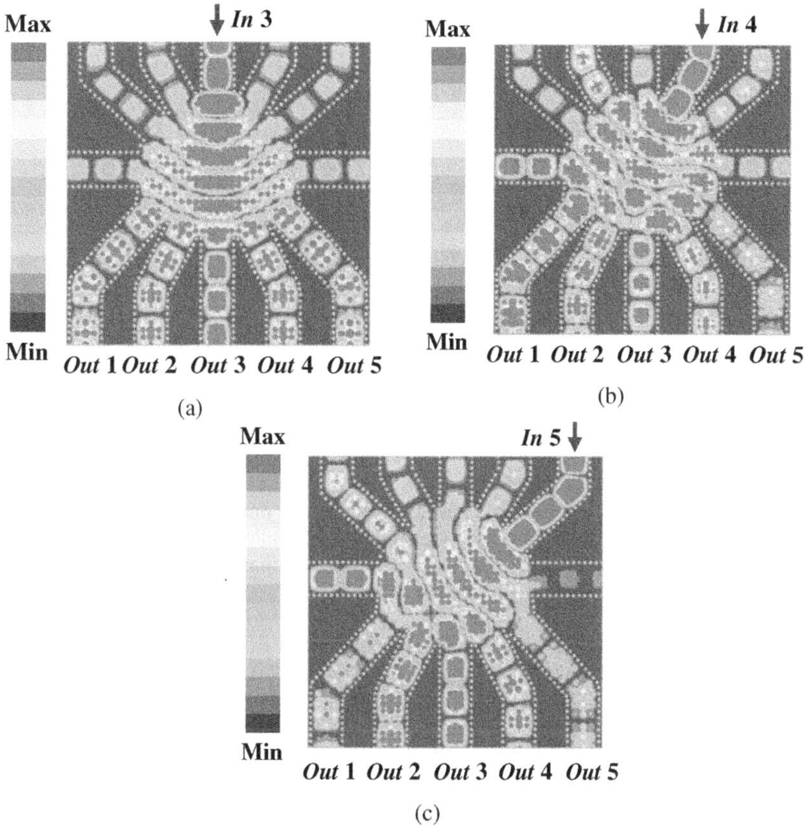

Figure 5.9 E-field magnitude distributions of the miniaturized SW-SIW Rotman lens fed by different input ports. (a) Fed by *In* 3. (b) Fed by *In* 4. (c) Fed by *In* 5.

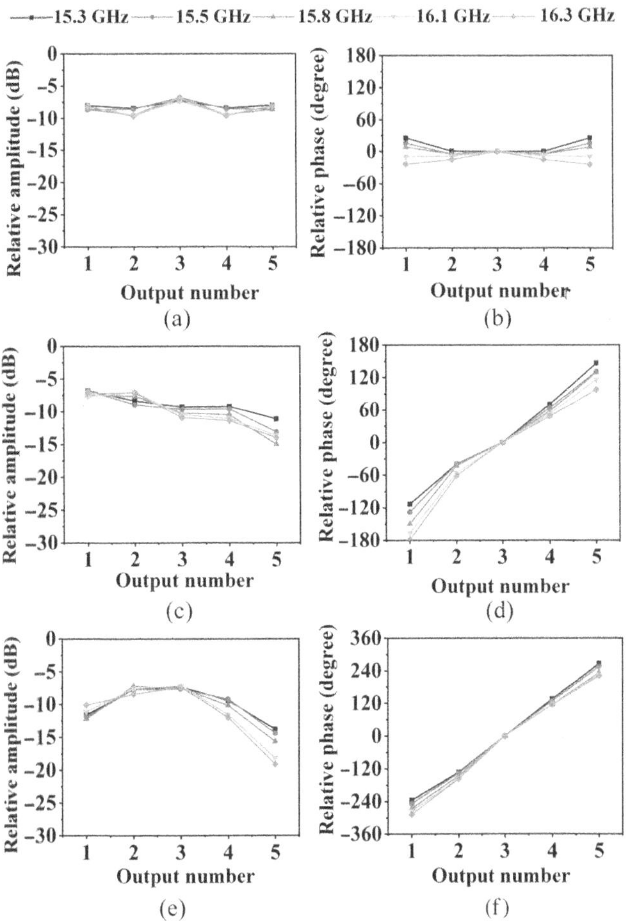

Figure 5.10 Relative amplitudes and phases of the output ports of the miniaturized SW-SIW Rotman lens. (a) Relative amplitudes fed by *In* 3. (b) Relative phase distributions fed by *In* 3. (c) Relative amplitudes fed by *In* 4. (d) Relative phases fed by *In* 4. (e) Relative amplitudes fed by *In* 5. (f) Relative phases fed by *In* 5.

The output amplitude and normalized phase of each output port supplied from input ports *In* 3 to *In* 5 are shown in Figure 5.10. Since the input ports of the Rotman lens are symmetrical with respect to port *In* 3, only examples fed by *In* 3, *In* 4, and *In* 5 are shown. The designed Rotman lens is found to offer outstanding amplitude and phase output characteristics across 15.3–16.3 GHz. Phase distribution at the outports of the small SW-SIW Rotman lens is essentially linear. The relatively constant phase differences between adjacent outports guarantee consistent and stable beam directions.

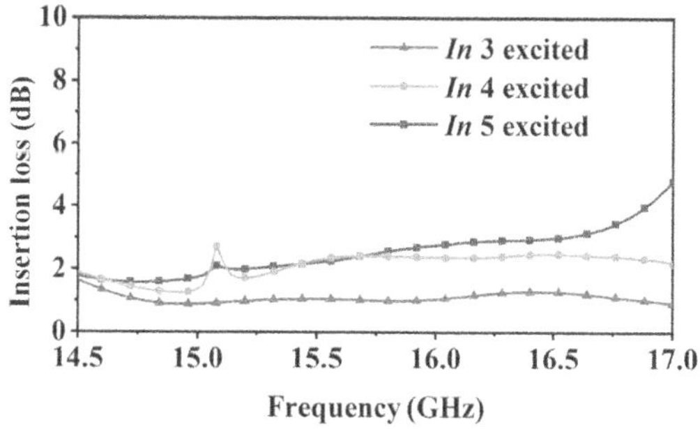

Figure 5.11 Insertion losses of beam ports in 3–5.

The insertion losses from the beam ports to the array ports are displayed in Figure 5.11. The proposed Rotman lens exhibits insertion losses of less than 1.3, 2.4, and 2.9 dB in the 15.3–16.3-GHz band, respectively, when *In* 3–*In* 5 are stimulated.

5.1.3 Design of Multibeam Antenna Based on the Miniaturized SW-SIW Rotman Lens

Evaluating the performance and feasibility of the multibeam antenna design based on the small SW-SIW Rotman lens is the aim. As illustrated in Figure 5.12, printed dipoles are employed as radiating elements and connected to the small SW-SIW Rotman lens's phase shifters (*Out* 1–*Out* 5) outports. The two arms of each dipole are connected independently to the top and bottom metal surfaces of the SW-SIW to finish the antenna system integration. Half-wavelength lengths are used to design the dipoles, which are subsequently adjusted to a final size of $L_3 = 4.7$ mm. The dipole La's height in relation to the substrate metal is adjusted for directional radiations at $L_a = 3$ mm. A tapered double strip line with terminal widths of $W_f = 2.2$ mm and $W_3 = 1$ mm is utilized as an impedance transformer to ensure sufficient impedance matching for each dipole on the operating band.

To enable measurements during the transition from SW-SIW to coaxial connections, tapered grounded coplanar waveguide (GCPW) transitions with one terminal width $W_1 = 1.85$ mm and the other terminal width $W_2 = 3.3$ mm are employed as impedance transformers. The geometrical dimensions of the multibeam antenna are listed in Table 5.1.

5.1.4 Fabrication and Measurement of the Multibeam Antenna

The small SW-SIW Rotman-lens-based multibeam antenna with printed dipoles as radiating elements has been successfully fabricated and measured. Images

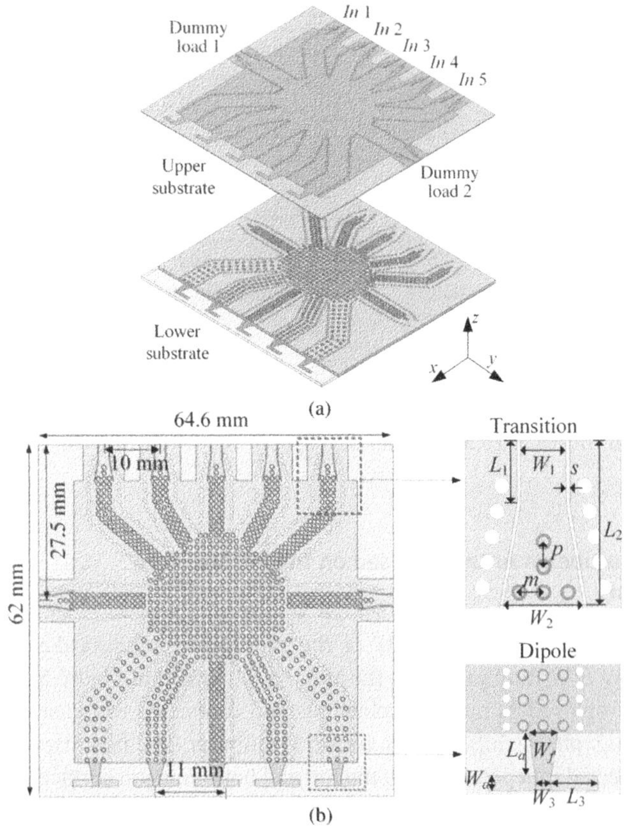

Figure 5.12 Configuration of the multibeam antenna based on miniaturized SW-SIW Rotman lens. (a) Exploded 3D view. (b) Top view.

Table 5.1 Final Parameters of the Designed Antenna.

Parameter	Value (mm)	Parameter	Value (mm)	Parameter	Value (mm)
L_1	2.5	W_1	1.85	s	0.15
L_2	6.5	W_2	3.3	m	1
p	1	W_3	1	L_3	4.7
W_1	1.1	W_f	2.2	L_a	3

Figure 5.13 Photographs of the fabricated multibeam antenna based on a miniaturized SW-SIW Rotman lens. (a) Top view. (b) Bottom view.

of the produced multibeam antennas are displayed in Figure 5.13. Additionally, Figure 5.14 demonstrates that in the 15.3–16.3-GHz frequency band, the reflection coefficients of the input ports of the multibeam antenna are less than −10 dB. Any discrepancies between the simulated and measured results are due to fabrication tolerances. Every part of the multibeam antenna, such as the input transitions, SW feeding lines, SW Rotman lens, SW phase shifters, and dipoles, will cause variations in models and measurements of the reflection coefficients.

Figure 5.15 confirms the beam-forming performance of the tiny SW-SIW Rotman lens, showing good agreement between the simulations and the observed multibeam directions. At the center frequency of 15.8 GHz, the equivalent actual gain values for the five simulated beam directions—−35°, −17°, 0°, 17°, and 34°—are 6.09, 7.71, 9.34, 8.02, and 6.32 dBi, respectively. The maximum directions of the five measured beams are −34.2°, −17.4°, 0°, 17.8°, and 35.5°, respectively, and their realized gain values are 4.29, 6.46, 8.06, 6.65, and 4.82 dBi, respectively. The small beam direction changes seen are caused by the output phase errors of the manufactured prototype, which are mostly the consequence of manufacturing and assembly flaws.

A comparison is made in Table 5.2 to show how the lens body size has decreased and how the entire Rotman lens of the proposed new design has become smaller. Despite the simple reduction of the feeding lines and delay lines in References [3–5], and, their lens bodies, which are crucial parts of beamforming networks, nevertheless occupied a large amount of area. In References [6, 7], Rotman lenses were reduced in size by employing coiled or folded designs. According to this design suggestion, a Rotman lens might be folded along the center plane [6] or wrapped with a special conformal substrate [7]. Although both folded designs were successfully 50% smaller, manufacturing problems prevented any

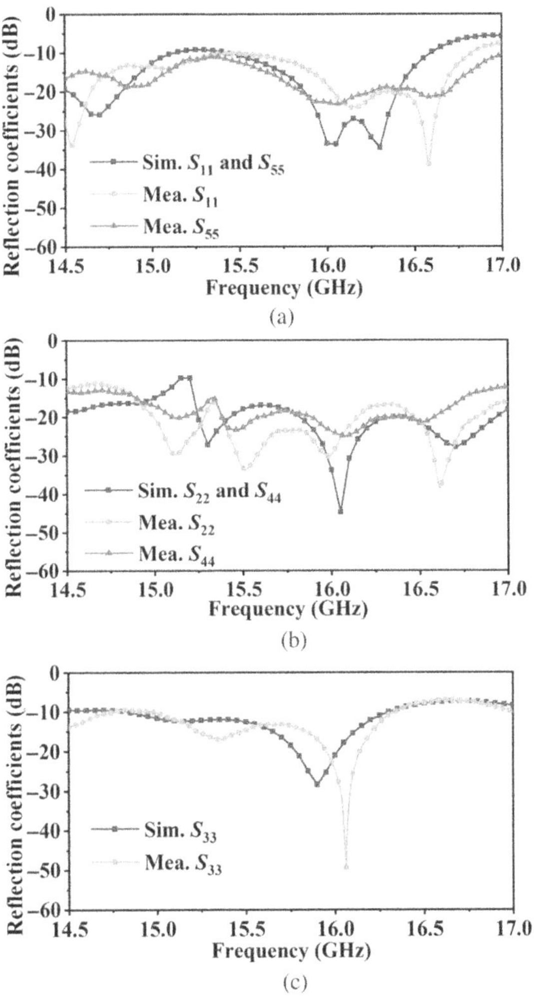

Figure 5.14 Simulated and measured reflection coefficients of the input ports of the multibeam antenna. (a) Fed by *In* 1 and *In* 5. (b) Fed by *In* 2 and *In* 4. (c) Fed by *In* 3.

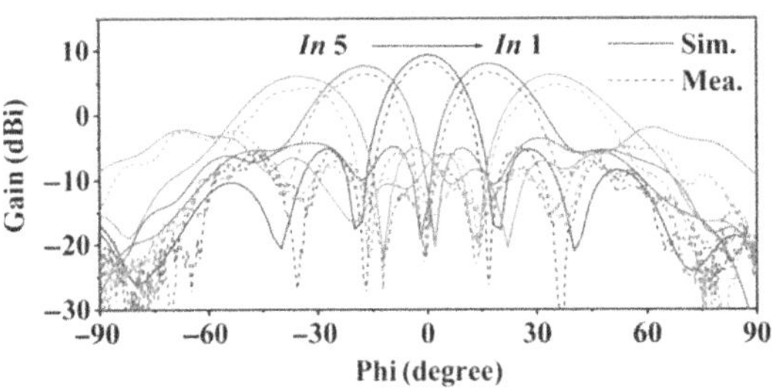

Figure 5.15 Radiation patterns of the multibeam antenna at 15.8 GHz.

Table 5.2 Comparisons with the Existing Miniaturized Rotman Lenses.

Reference	Transmission line type	Area[a)] ($\lambda_0 \times \lambda_0$)	Beam scanning range	Maximum insertion loss	Miniaturized part	Miniaturization technologies	Size reduction of the lens body	Fabrication difficulty
[3]	Microstrip	$9.6\lambda_0 \times 11.6\lambda_0$	−30° to 30°	>6 dB	Delay lines	Chebyshev impedance transformers	None	Easy
[4]	Microstrip	$9.2\lambda_0 \times 6.3\lambda_0$	−33° to 33°	>5 dB	Feeding lines and delay lines	Truncated ports with energy distribution slots	None	Easy
[5]	Microstrip	$9.2\lambda_0 \times 6.3\lambda_0$	−33° to 33°	>3 dB	Feeding lines	Power dividers connected to beam ports	None	Easy
[6]	Microstrip	Not given	−30° to 30°	Not given	Lens body	Folded configuration	50%	Difficult
[7]	Microstrip	$10.3\lambda_0 \times 6.0\lambda_0$	−30° to 30°	Not given	Lens body	Wrapped configuration using a special conformal substrate	50%	Very difficult
This work	Substrate-integrated waveguide	$2.6\lambda_0 \times 2.7\lambda_0$	−35° to 35°	2.9 dB	Lens body	Slow-wave substrate-integrated waveguide	66.1%	Easy

a) There are areas of the Rotman lenses that include the input feeding lines and output phase shifters.

quantitative results from being provided to support the feasibility of the design concept. Furthermore, the input feeding lines and output phase shifters are designed using the SW-SIW, which leads to significantly shorter lengths than those of conventional SIW-based ones. The electrical dimensions of the proposed SW-SIW Rotman lens, including the input feeding lines and output phase shifters, are therefore far smaller than those of the existing micro-Rotman lenses. In conclusion, the SW-SIW Rotman lens developed in this work successfully achieves a significant size reduction using widely accessible manufacturing technology. The proposed compact multibeam antenna is a perfect choice for satellite-assisted IoVs applications, based on the miniature SIW Rotman lens.

5.1.5 Conclusion

The study's findings, which center on the development and evaluation of a small multibeam antenna for IoV applications aided by Ku-band satellites, are shown in this part. A tiny SW-SIW Rotman lens serves as the basis for the antenna. A notable size decrease of 66.1% was achieved by integrating SW structures into the SW-SIW lens. A multibeam antenna prototype was made and tested using the SW-SIW Rotman lens in order to validate the design concept. The radiating components of the prototype were printed dipoles. The precision and dependability of the design are confirmed by the strong agreement between the measured beam directions and the anticipated outcomes. The prototype's reported beam directions were—34.2°, −17.4°, 0°, 17.8°, and 35.5°. According to these findings, the small multibeam antenna with the reduced SW-SIW Rotman lens is ideal for Ku-band frequency-based satellite-assisted IoV applications. The successful manufacture and testing of the prototype proved the viability and usefulness of the suggested design.

References

1 Cheng, Y.J., Hong, W., Wu, K. et al. (2008). Substrate integrated waveguide (SIW) Rotman lens and its Ka-band multibeam array antenna applications. *IEEE Transactions on Antennas and Propagation*, 56 (8), 2504–2513.

2 Wu, Y.F., Cheng, Y.J., and Huang, Z.X. (2018). Ka-band near-field-focused 2-D steering antenna array with a focused Rotman lens. *IEEE Transactions on Antennas and Propagation*, 66 (10), 5204–5213.

3 Liang, Q., Sun, B., Zhou, G., and Li, J. (2018). Compact microstrip Rotman lens using Chebyshev impedance transformers. *Progress in Electromagnetics Research Letters*, 76, 1–6.

4 Liang, Q., Sun, B., Zhou, G., Zhao, J., and Zhang, G. (2019). Design of compact Rotman lens using truncated ports with energy distribution slots. *IEEE Access*, 7, 120766–120773.

5 Sun, B., Liang, Q., and Zhou, G. (2020). Miniaturized Rotman lens with applications to wireless communication. *Frontiers of Information Technology & Electronic Engineering*, 21 (1), 144–158.

6 Vo Dai, K. T., Nguyen, T., and Kilic, O. (2017). A compact microstrip Rotman lens design. in *Proceedings of the United States National Committee URSI National Radio Science Meeting (USNC-URSI NRSM)* (January 2017), pp. 1–2.

7 Dai, K.T.V. and Kilic, O. (2016). Compact Rotman lens structure configurations to support millimeter wave devices. *Progress in Electromagnetics Research B*, 71, 91–106.

8 Lee, W., Kim, J., and Yoon, Y.J. (2011). Compact two-layer Rotman lens-fed microstrip antenna array at 24 GHz. *IEEE Transactions on Antennas and Propagation*, 59 (2), 460–466.

9 Tekkouk, K., Ettorre, M., Coq, L.L., and Sauleau, R. (2016). Multibeam SIW slotted waveguide antenna system fed by a compact dual-layer Rotman lens. *IEEE Transactions on Antennas and Propagation*, 64 (2), 504–514.

10 Niembro-Martínet, A., Nasserddine, V., Pistono, E. et al. (2014). Slow-wave substrate integrated waveguide. *IEEE Transactions on Microwave Theory and Techniques*, 62 (8), 1625–1633.

11 Jin, H.-Y., Wang, K.-D., Guo, J.-P., Ding, S., and Wu, K. (2016). Slowwave effect of substrate integrated waveguide patterned with microstrip polyline. *IEEE Transactions on Microwave Theory and Techniques*, 64 (6), 1717–1726.

12 Zhang, Y., Deng, J.Y., Sun, D. et al. (2021). Slow wave substrate-integrated waveguide with miniaturized dimensions and broadened bandwidth. *IEEE Transactions on Microwave Theory and Techniques*, 69 (8), 3675–3683.

13 Bertrand, M., Liu, Z., Pistono, E., Kaddourt, D., and Ferrar, P. (2015). A compact slow-wave substrate integrated waveguide cavity filter. in *IEEE MTT-S International Microwave Symposium*, Phoenix, AZ, USA (May 2015), pp. 1–3.

14 Guan, D.-F., Zhang, Q., You, P., Yang, Z.-B., Zhou, Y., and Yong, S.-W. (2018). Scanning rate enhancement of leaky-wave antennas using slow-wave substrate integrated waveguide structure. *IEEE Transactions on Antennas and Propagation*, 66 (7), 3747–3751.

15 Ho, A.T., Pistono, E., Corrao, N., and Ferrari, P. (2021). Circular polarized square slot antenna based on slow-wave substrate integrated waveguide. *IEEE Transactions on Antennas and Propagation*, 69 (3), 1273–1282.

6

Conclusion

For wireless communication systems to be more efficient in terms of spectrum, energy, and cost, miniaturized and integrated multi-beam antennas utilizing slow-wave substrate-integrated waveguide (SW-SIW) are essential. The study and analysis of SW-SIW transmission lines, devices, beamforming networks, and multi-beam antennas are thoroughly reviewed and described in this book. Since multi-beam antennas may produce several high-gain directed beams, increasing the communication system's capacity, their performance is essential to the overall wireless communication system. The SIW, which combines the benefits of metallic waveguides and microstrip lines with low cost, low loss, and simple integration with planar circuits, is perfect for designing multi-beam antennas in light of the trend toward miniaturization and integration in wireless communication systems. Additionally, it reduces the size of antennas and microwave devices by shortening the waveguide wavelength.

The SW-SIWs in integrated and miniature antennas are the primary topic of this book. In addition to covering the basics of SW-SIW transmission lines, it also provided information on various device and antenna types and showed the design and working principle of the SW-SIW. Practical examples of several SW-SIWs are included in this book to help readers get acquainted with them and quickly get started with practical antenna design techniques. These real-world examples greatly aid readers in quickly understanding and effectively applying the information in this specialist field.

Slow-Wave Substrate-Integrated Waveguide: Designing of Devices and Antennas, First Edition.
Jing-Ya Deng, Jia-Yuan Yin, Fengxia Li, Yulong Liu, Li-Xin Guo, and Xiao-Hua Ma.

Index

Slow-Wave Substrate-Integrated Waveguide: Designing of Devices and Antennas, First Edition.
Jing-Ya Deng, Jia-Yuan Yin, Fengxia Li, Yulong Liu, Li-Xin Guo, and Xiao-Hua Ma.

t

u

v

w

Printed and bound by CPI Group (UK) Ltd, Croydon, CR0 4YY

17/09/2026

14964232-0001